Smart Aquaponic Farming

An Intelligent Production System for Aquaponics

Smart Aquaponic Farming

An Intelligent Production System for Aquaponics

Edited by

Daoliang Li

National Innovation Center for Digital Fishery, China Agricultural University, Beijing, P.R. China; College of Information and Electrical Engineering, China Agricultural University, Beijing, P.R. China

ELSEVIER

ACADEMIC PRESS

An imprint of Elsevier

Academic Press is an imprint of Elsevier
125 London Wall, London EC2Y 5AS, United Kingdom
50 Hampshire Street, 5th Floor, Cambridge, MA 02139, United States

ISBN: 978-0-443-28945-3

For Information on all Academic Press publications
visit our website at https://www.elsevier.com/books-and-journals

Publisher: Mara Conner
Acquisitions Editor: Vaishali Singh
Editorial Project Manager: Gulafshan Iqbal
Production Project Manager: Fahmida Sultana
Cover Designer: Arumugam Kothandan

Typeset by MPS Limited, Chennai, India

Contents

List of contributors

Zhuangzhuang Bai
National Innovation Center for Digital Fishery, China Agricultural University, Beijing, P.R. China; College of Information and Electrical Engineering, China Agricultural University, Beijing, P.R. China

Haojie Chang
Centre for Vision Speech and Signal Processing, University of Surrey, Guildford, United Kingdom

Yaru Chen
Centre for Vision Speech and Signal Processing, University of Surrey, Guildford, United Kingdom

Meng Cui
Centre for Vision Speech and Signal Processing, University of Surrey, Guildford, United Kingdom

Zhuangzhuang Du
National Innovation Center for Digital Fishery, China Agricultural University, Beijing, P.R. China; School of Computer and Information Engineering, Henan University, Kaifeng, P.R. China

Quanwu Ge
National Innovation Center for Digital Fishery, China Agricultural University, Beijing, P.R. China; College of Information and Electrical Engineering, China Agricultural University, Beijing, P.R. China

Youbang Guan
Centre for Environment and Sustainability, School of Sustainability, Civil and Environmental Engineering, University of Surrey, Guildford, United Kingdom

Jie Han
National Innovation Center for Digital Fishery, China Agricultural University, Beijing, P.R. China; College of Information and Electrical Engineering, China Agricultural University, Beijing, P.R. China

Lingwei Jiang
National Innovation Center for Digital Fishery, China Agricultural University, Beijing, P.R. China; School of Chemistry and Chemical Engineering, University of Surrey, Guildford, United Kingdom

Daoliang Li
National Innovation Center for Digital Fishery, China Agricultural University, Beijing, P.R. China; College of Information and Electrical Engineering, China Agricultural University, Beijing, P.R. China

Chang Liu
National Innovation Center for Digital Fishery, China Agricultural University, Beijing, P.R. China

Lihua Song
School of Chemistry & Chemical Engineering, University of Surrey, Guildford, United Kingdom

Bingxiong Wang
National Innovation Center for Digital Fishery, China Agricultural University, Beijing, P.R. China; College of Information and Electrical Engineering, China Agricultural University, Beijing, P.R. China

Cong Wang
National Innovation Center for Digital Fishery, China Agricultural University, Beijing, P.R. China

GuangXu Wang
National Innovation Center for Digital Fishery, China Agricultural University, Beijing, P.R. China; College of Information and Electrical Engineering, China Agricultural University, Beijing, P.R. China

Liang Wang
National Innovation Center for Digital Fishery, China Agricultural University, Beijing, P.R. China; School of Chemistry and Chemical Engineering, University of Surrey, Guildford, United Kingdom

Xiaoyang Wu
School of Chemistry and Chemical Engineering, University of Surrey, Guildford, United Kingdom

Xianbao Xu
College of Information and Electrical Engineering, China Agricultural University, Beijing, Beijing, P.R. China; College of Electronics and Electrical Engineering, Anhui Agricultural University, Hefei, Anhui, P.R. China

Jinqi Yang
School of Chemistry and Chemical Engineering, University of Surrey, Guildford, United Kingdom

Pan Zhang
National Innovation Center for Digital Fishery, China Agricultural University, Beijing, P.R. China; College of Information and Electrical Engineering, China Agricultural University, Beijing, P.R. China

About the editor

Daoliang Li, PhD, is the dean of the International College Beijing and a professor and doctoral supervisor of the College of Information and Electrical Engineering at China Agricultural University. He is a distinguished professor, director of the National Innovation Center for Digital Fishery, and deputy chairman of the Agricultural and Rural Informatization Expert Advisory Committee of the Ministry of Agriculture and Rural Affairs. He has long been engaged in intelligent agricultural information processing and agricultural and rural informatization research. He has successively undertaken 108 national and provincial-level projects, including the National Natural Science Foundation, the National 863 Program, the National Science and Technology Support Program, the National Key Research and Development Program, the Seventh Framework and Horizon 2020 Program of the European Union, and the National Major Science and Technology Research Program. He has been selected as the winner of the National Ten-Thousand Program, the Young and Middle-Aged Leading Talents of the Ministry of Science and Technology, the Shennong Leading Talents of the Ministry of Agriculture and Rural Affairs, outstanding scientific research talents, and the New Century Talents of the Ministry of Education. He was also the winner of the National Innovation Prize, the China Youth Science and Technology Award, the Outstanding Engineer Award of the China Foundation for International Science Exchange, the National Outstanding Scientific and Technological Worker, and the Outstanding Communist Party Member of Beijing Universities. He is currently the chairman of the Agricultural Advanced Information Processing Division of the International Federation for Information Processing (IFIP), editor-in-chief of the international journal *Information Processing in Agriculture*, chief scientist of the Hebei Province Freshwater Aquaculture Industry Technology System, and post-intelligent aquaculture scientist of the National Shrimp and Crab Industry Technology System. He has won 1 second prize of the National Science and Technology Progress Award, 4 first prizes of the Provincial and Ministerial Science and Technology Progress Award, and 61 authorized national invention patents. He has published 185 SCI-indexed papers as the first author or corresponding author and 20 monographs, such as *Unmanned Farms: A New Model for Future Agriculture*, *Agriculture 4.0: The Coming Era of Smart Agriculture*, *Internet + Agriculture: The Inevitable Path of Agricultural Supply-Side Reform*, and formulated 2 national standards and 7 industrial and local standards. The team he led was selected as the innovation team of the Ministry of Agriculture and Rural Affairs.

Preface

Aquaponics is a novel integrated farming system that achieves a scientific synergistic symbiosis between fish, vegetables, and microorganisms through clever ecological design of aquaculture and hydroponics. This results in fish farming without water changes and water quality concerns and vegetables growing normally without fertilizers, creating a mutually beneficial ecological effect. Industrialized aquaponics systems integrate recirculating aquaculture systems and hydroponics plants on a large scale. Utilizing advanced information communication technology, biotechnology, artificial intelligence (AI), and embodied intelligence, it achieves dynamic balance of matter, energy, and gases in production, and optimal scheduling of production materials, personnel, and equipment in management. This modern agricultural production paradigm features resource recycling, high input and high output, high production efficiency, and significant low-carbon ecological benefits. Smart aquaponics farming represents the pinnacle of industrialized aquaponics systems and is becoming increasingly popular and developed in countries and regions facing resource and population constraints and urgent needs for modern agricultural transformation and upgrading.

Our team is one of the earliest in China to engage in scientific research and industrial practice of industrialized aquaponics technology. From 2012 to 2014, we achieved version 1.0 of a direct-connection, reusable aquaponics production greenhouse in Shouguang, Shandong Province, China. From 2014 to 2016, we achieved version 2.0 of a large-scale bypass reusable aquaponics production plant in Wuqing, Tianjin, China. From 2016 to 2019, we achieved version 3.0 of a two-main-one-auxiliary aquaponics intelligent factory in Jiulongpo, Chongqing, China. From 2020 to 2024, we achieved version 4.0 of a data-driven full-chain aquaponics intelligent factory in Guangzhou, Guangdong Province, China. Starting in 2024, we have been promoting the iteration of version 5.0 of an AI-model-driven autonomous aquaponics intelligent factory in Liangping, Chongqing, China. Throughout our practical development, the technological bottlenecks in smart aquaponics farming still lie in fundamental contradictions in the needs of aquatic animals and plants, the craft requirements involving multiple links and factors, the lack of intelligent control models that integrate multi-target recognition and multimodal fusion, and the absence of autonomous intelligent equipment adaptable to multiple scenarios and tasks. At the scientific research level, breakthroughs are still needed in areas such as the digital perception of large-scale resource, environmental, biological, microbial, and equipment information; networked transmission from cloud to network to edge to terminal; digitalization and mechanistic modeling of integrated processes; cleaning, fusion, feature engineering, and knowledge graphing of multisource heterogeneous data; large-scale multimodal fusion of individual and group intelligent recognition and decision-making models; deployment of small models on the edge after distillation or pruning; and intelligent aquaculture equipment adaptable to all tasks.

This book systematically reviews and presents the development history, system composition, core technology analysis and implementation, and global case studies of smart aquaponics farming from an industry perspective. It aims to inspire researchers to enrich and develop smart aquaponics farming technology and help practitioners implement and improve smart aquaponics factory production. This book can also serve as a reference for engineering and technical personnel and a textbook for students in related majors at universities and colleges.

This book is organized into 11 chapters: Chapter 1 thoroughly provides an overview of the concept and scope of smart aquaponics farming and its chapter arrangement. Chapters 2–10 systematically describe the various components of smart aquaponics farming and core topics such as air environment control, precision feeding, equipment energy consumption management, microbial regulation, system integration, and sustainable analysis and evaluation. Chapter 11 mainly selects and presents representative aquaponics production system cases from around the world.

In detail, Chapter 1, "Introduction," focuses on the history, development, and technological evolution of aquaponics, presenting the overall concept, characteristics, advantages, and future challenges of smart aquaponics farming, along with a description of all chapters in the book.

Chapter 2, "Digital infrastructure system," mainly introduces and explains the important contents of infrastructure foundation, as well as the methods and significance of realizing digital monitoring of infrastructure system.

Chapter 3, "Recirculating aquaculture system," will focus on the fish farming component, explaining the principles of recirculating water technology, intelligent water quality monitoring and regulation, and how to achieve extreme water resource recycling.

Chapter 4, " Hydroponic planting system," will explore plant cultivation techniques, including different soilless cultivation models, intelligent management of nutrient solutions, and optimization strategies for plant growth environments.

Chapter 5, "air environment control system," will deeply monitor units for spatiotemporal data of greenhouse external and internal climate, as well as monitor units for internal air environment parameters, microclimate models, and environmental parameter optimization and control technologies.

Chapter 6, " Fish precision feeding system," will explain how to achieve precise feeding based on fish growth stages and feeding demands through intelligent equipment and algorithms, maximizing feed utilization and reducing water pollution.

Chapter 7, "Equipment and energy consumption control system," will introduce key production equipment within the factory and focus on how to optimize energy efficiency through intelligent means, integrate renewable energy, and perform predictive maintenance to ensure stable system operation.

Chapter 8, "Microbial regulatory system," will reveal the critical role of microbial communities in aquaponics systems. From the aspects of bacteria, fungi, viruses, and microalgae, the threats posed by pathogenic microorganisms are sorted out and described, as well as their monitoring and prevention measures.

Chapter 9, "Smart aquaponics farming integration," will elaborate on how intelligent aquaponics farming integration for intelligent operation through sensor networks, data acquisition and transmission, and big data cloud platforms driven by data and AI models achieves significant overall efficiency improvements and system optimization through intelligent integrated control.

Chapter 10, "Sustainability evaluation of smart aquaponics farming," will establish a comprehensive assessment framework to quantify the economic, net-zero, ecological, and social benefits of aquaponics farming.

Chapter 11, "Case studies of aquaponics systems worldwide," will showcase successful practices of advanced aquaponics farming through international cases, select representative aquaponics farming from countries on various continents, introduce their basic situation, and summarize the system characteristics and experience practices.

I wish to thank many colleagues and PhD students who were involved in contributing to this book: Liang Wang (Chapters 1 and 2); Cong Wang, Lingwei Jiang, and Chang Liu (Chapter 3); Pan Zhang, Lingwei Jiang, and Jinqi Yang (Chapter 4); Xiaoyang Wu and Jinqi Yang (Chapter 5); Guangxu Wang and Zhuangzhuang Du (Chapter 6); Bingxiong Wang and Quanwu Ge (Chapter 7); Zhuangzhuang Bai and Jie Han (Chapter 8); Liang Wang and Xianbao Xu (Chapter 9); Youbang Guan and Lihua Song (Chapter 10); Meng Cui, Haojie Chang, and Yaru Chen (Chapter 11). We are grateful to those at Elsevier Global Book Production, with a special acknowledgment to Fahmida Sultana (Project Manager), Kyle Gravel, Gulafshan Iqbal, and Jeraldine Calagui (Content Manager), Dinesh N. and Varshini V. (Copyright Coordinator), and Goutham Sampathkumar (Contracts Coordinator). We would like to express our gratitude to the Agricultural Engineering Research Institute of the Chongqing Academy of Agricultural Sciences, our collaborative partner in research and industrial practice, for their support and assistance. Special thanks go to Lihong Gao (Director), Ping Li, Jishu Zheng (Deputy Director), and the engineers in their team. We are also grateful to the National Natural Science Foundation of China (Grant No. 3237318, Dynamic Regulation Mechanism of Nitrogen in Industrial Aquaponics under the Condition of Asynchronous Life Cycle) and the China Scholarship Council (Cross-Innovation Talent Training Program for Digital Fishery) for providing research and training support for this book.

Prof. Dr. Daoliang Li

CHAPTER

Introduction

1

Daoliang Li[1,2] and Liang Wang[1,3]
[1]National Innovation Center for Digital Fishery, China Agricultural University, Beijing, P.R. China,
[2]College of Information and Electrical Engineering, China Agricultural University, Beijing, P.R. China,
[3]School of Chemistry and Chemical Engineering, University of Surrey, Guildford, United Kingdom

1.1 Aquaponics: a fusion of tradition and modernity

1.1.1 Concepts and development background: from ancient wisdom to modern agricultural evolution

Aquaponics, as a novel integrated agricultural system, ingeniously combines two independent production technologies: aquaculture and hydroponics, creating a mutually beneficial circular ecosystem (Goddek et al., 2015). Its core principle lies in the fact that fish waste, after microbial conversion, provides essential nutrients for plants, while plants, through their root systems, absorb these nutrients and simultaneously purify the water, allowing clean water to return to the fish tanks for fish growth. This ingenious ecological design achieves the synergistic effect of "raising fish without changing water and without water quality concerns, and growing vegetables without fertilization while growing normally."

Aquaponics is not a modern agricultural invention. Its early forms can be traced back to the ancient agricultural practices of the Aztec *chinampas* and Chinese *Mulberry/Rice-Fish Pond*, which already understood how to utilize aquatic environments for synergistic cultivation and aquaculture (Somerville et al., 2014). With the advancement of modern technology, particularly the maturity and development of recirculating aquaculture systems and hydroponic techniques, aquaponics systems have achieved a qualitative leap based on scientific theory and engineering practice. Traditional agricultural models worldwide are facing increasingly severe challenges, including water scarcity, land degradation, and pollution caused by excessive use of chemical fertilizers and pesticides (FAO, 2020). In this context, aquaponics emerges as a sustainable solution, offering new perspectives and avenues to address these challenges (Love et al., 2015).

Currently, continuous global population growth, accelerated urbanization, climate change, and increasingly prominent food security issues highlight the limitations of traditional agriculture. Aquaponics, as an intensive, efficient, and environmentally

Smart Aquaponic Farming. DOI: https://doi.org/10.1016/B978-0-443-28945-3.00001-3

friendly agricultural production method, is considered a key technology for enhancing food security and ensuring the supply of high-quality agricultural products. This model can achieve high-density aquaculture and cultivation under limited land and water resources, especially suitable for urban agriculture and water-scarce regions, effectively alleviating resource pressure. Aquaponics is a comprehensive solution to global food and environmental challenges, realizing its potential through optimization and integration of existing principles, rather than creating something out of nothing. For example, the prominent position of aquaponics in urban agriculture is not only due to its production efficiency but also its diversified potential for social and economic development. It can be deeply integrated with popular science education, tourism, and other industries, attracting urban and rural consumers for sightseeing and shopping, creating new commercial selling points, thereby broadening the traditional boundaries of agriculture and achieving multiple benefits.

1.1.2 Types and processes development: from simple small size to multiple commercial scale

The emergence of modern aquaponics in the 1970s is largely attributed to the pioneering work of John and Nancy Todd and William McLarney at the New Alchemy Institute. This was further advanced by Dr. Mark McMurtry's team at North Carolina State, whose reciprocating technologies significantly improved the system's efficiency (Delaide et al., 2016). By the 1990s, aquaponics technologies had matured through extensive research and practical application, setting the stage for further development. The 2000s saw the rise of large-scale and specialized aquaponics, including indoor, urban, and educational applications, which have since led to breakthrough commercial developments.

Aquaponics systems can be categorized based on their intermediate processing and transformation stages:

- **Open-loop aquaponics:** In this design, effluent from fish tanks is used for plant irrigation and nutrition but does not return to the tanks.
- **Closed-loop aquaponics:** This system processes fish tank effluent, uses it to nourish plants, and then recirculates the purified water back to the fish tanks. This closed-loop approach is highly water-efficient, significantly reducing water consumption compared to traditional methods (Delaide et al., 2016). Advanced closed-loop systems may also incorporate bypass pipe systems for balanced water dynamics and one-way condensation systems for horticulture, ensuring ideal growth conditions for both fish and vegetables.

In terms of application, aquaponics systems serve diverse purposes:

- **Commercial aquaponics:** Focuses on large-scale agricultural production of high-value fish, plants, algae, and fungi, with strict consideration for economic viability and market demands (Palm et al., 2018).

- **Family/living aquaponics:** Includes backyard, indoor, or even desktop setups designed for household use, producing fresh, organic fish, vegetables, fruits, and flowers for personal consumption and leisure.
- **Educational aquaponics:** Primarily used for demonstrating aquaponics principles and serving as educational models.

Regarding the growing environment, systems can be classified as:

- **Open aquaponics:** Systems exposed to the natural environment.
- **Greenhouse aquaponics:** These systems leverage facility greenhouse technologies and intelligent control systems to enable automated, year-round, sustained yields.

The adaptable nature of aquaponics allows for significant structural modifications to suit specific local and worksite conditions, such as landform, restricted space, water sources, climate, and economic viability. This flexibility has greatly enriched the design and application possibilities of aquaponics systems.

1.1.3 The evolution of intelligence: Internet of Things and artificial intelligence empowering future agriculture

Traditional aquaponics systems are often time-consuming and labor-intensive to manage, making it difficult to maintain optimal operating conditions. They also lack real-time monitoring and automated regulation mechanisms, leading to problems such as water quality fluctuations and nutrient imbalances, which in turn affect fish and plant growth and system stability (Karimanzira and Jacobi, 2021). This over-reliance on manual experience limits their scalability and efficiency.

The integration of Internet of Things (IoT) technology has brought revolutionary changes to aquaponics systems. By deploying various stable and easy-to-maintain sensors and monitoring equipment, intelligent systems can perform real-time, stable, and continuous monitoring and data collection of key parameters in the aquaponics environment, such as temperature, humidity, light, carbon dioxide, dissolved oxygen, pH, electrical conductivity, turbidity, liquid level, and flow rate (Kyaw and Ng, 2017). This massive amount of data is transmitted wirelessly to a central control unit or cloud platform, laying a solid foundation for intelligent management.

Artificial intelligence (AI) technology plays the role of the "brain" in intelligent aquaponics farming, elevating the system from a biological system to a data-driven intelligent ecosystem. Through in-depth analysis of real-time and historical data, AI algorithms can: predict system behavior, such as fish growth trends, plant growth patterns, and nutrient requirements; enable automatic early warning and auxiliary decision-making (Mahon et al., 2022), where if sensors detect abnormal conditions (e.g., water quality parameter deviations, signs of equipment malfunction), the AI system can automatically issue warnings and provide corrective action suggestions, or even automatically activate corresponding equipment for adjustment; optimize resource management, where AI algorithms can precisely control environmental parameters such as water flow, aeration, nutrient solution dosing, light intensity, and temperature, ensuring optimal growth conditions for fish and plants while achieving precise and economical use of water, energy, feed, and other resources. Furthermore, combined with mobile applications and web platforms, users can remotely view system status, receive abnormal alarms, and perform remote control anytime, anywhere.

The introduction of intelligence significantly enhances production precision, scientific management, efficiency, and quality control capabilities, reducing manual intervention and labor intensity, lowering operational risks and costs, and improving product quality and traceability. This paradigm shift from "experience-based management" to "precise prediction and control" pushes aquaponics systems from "semi-automation" toward "unmanned" production models. The combination of IoT and AI enables the system to operate efficiently as an organic whole, achieving closed-loop intelligent management from "sensing-analysis-decision-execution," which is its fundamental difference from traditional models. Smart aquaponics farming is not only an agricultural production revolution but also a typical example of the deep integration of cutting-edge technologies such as digital twins, big data, and cloud computing in agriculture, foreshadowing the "factory-like" and "unmanned" trend of future agriculture.

1.2 Overview of smart aquaponics farming

1.2.1 Definition and core characteristics

An smart aquaponics factory refers to a new agricultural production model in a controlled environment that deeply integrates advanced technologies (Wright et al., 2022) such as the IoT, AI, big data analytics, and automation control to achieve efficient synergy between fish farming and soilless plant cultivation, resource recycling, precise environmental parameter regulation, and intelligent management of the production process (Li et al., 2020). It goes beyond traditional aquaponics, pushing agricultural production toward a new stage of industrialization, intensification, and precision through data-driven and automation.

The core characteristics of smart aquaponics farming include:

- **High integration:** The system organically integrates multiple subsystems such as aquaculture, soilless cultivation, water treatment, environmental control, and energy management, forming a unified, collaborative production entity.
- **Intelligence and automation:** Extensive application of sensors, controllers, actuators, AI algorithms, and machine learning models enables real-time monitoring, prediction, automatic adjustment, and remote control of environmental parameters, significantly reducing manual intervention.
- **Efficient resource recycling:** Water utilization efficiency is extremely high, saving 90% or even 95% of water (Graber and Junge, 2009); fish waste, after microbial conversion, serves as plant nutrients, achieving zero nutrient discharge and zero chemical fertilizer use; energy consumption is effectively controlled through optimized management and renewable energy integration.

- **Environmental friendliness:** The system does not use chemical fertilizers, pesticides, or herbicides, reducing agricultural non-point source pollution and achieving a "zero-emission" new model for fisheries aquaculture.
- **High yield, high quality, and year-round production:** In a controlled environment, crop growth cycles are shortened, yields are significantly increased, and production is not limited by seasons or climate, enabling continuous year-round production with higher quality and safer products.

The "high integration" and "intelligence" in the core characteristics are not isolated but mutually supportive. Intelligent technologies (IoT, AI) are the adhesive that enables seamless integration and collaborative operation of various subsystems, thereby elevating a simple physical combination to an efficient ecological factory. For example, traditional aquaponics is a combination of aquaculture and hydroponics, but its management still requires significant manual intervention. Intelligent farming, however, digitize each link through sensors and controllers, and AI algorithms analyze this data to coordinate the operation of different systems (e.g., water quality, air, feeding), ultimately achieving overall optimization and efficiency improvement. This integrated control is the core manifestation of intelligence, enabling the system to operate efficiently as an organic whole.

1.2.2 Core advantages: resource efficiency, environmental benefits, and economic potential

The advantage of smart aquaponics farming lies in their multi-dimensional (Table 1.1), systemic benefits, rather than a single aspect of improvement (Khandakar et al., 2024).

1.2.2.1 Resource efficiency

- **Water saving:** Compared to traditional agriculture, aquaponics systems save up to 90% or even 95% of water. Water is recycled between fish tanks and plant grow beds, with only water lost due to natural evaporation and crop absorption needing to be replenished. Some advanced systems can even reduce water replenishment rates to 1%–3%.
- **Land saving:** High-density fish farming and high-density cultivation, especially vertical farming models, significantly increase output per unit area (Benis and Ferrão, 2017). This gives aquaponics a significant advantage in urban areas or arid regions with limited land resources.
- **Fertilizer saving:** Fish waste serves as natural fertilizer, eliminating the need for additional chemical fertilizers and reducing reliance on external nutrient sources (Wongkiew et al., 2017).

1.2.2.2 Environmental benefits

- **Zero emission and low carbon:** Intelligent aquaponics systems achieve "zero discharge" of aquaculture wastewater, avoiding pollution of the environment by

Table 1.1 Core advantages comparison of smart aquaponics farming.

Feature dimension	Traditional agriculture	Traditional aquaculture	Traditional hydroponics	Intelligent aquaponics factory
Water resource utilization rate	Low (High consumption)	Medium (Requires water change & discharge)	High (Requires water & nutrient replenishment)	Extremely High (90%–95% Water Saving)
Land use efficiency	Low (requires large arable land)	Medium (requires water area)	High (vertical)	Extremely high (high-density, vertical)
Chemical fertilizer and pesticide use	High	Medium (may use antibiotics)	High (requires chemical fertilizers)	Zero (fish waste provides nutrients, no pesticides)
Labor demand	High	Medium	Medium	Low (semi-automated/ unmanned)
Environmental pollution (wastewater/ carbon emissions)	High (non-point source pollution)	High (tailwater pollution)	Low (wastewater discharge)	Extremely low (zero emission, low carbon footprint)
Production cycle/yield	Seasonally limited, yield fluctuates	Long cycle, limited yield	High yield, but single product	Year-round production, high yield and value
Product quality/safety	Affected by soil/ environment	May contain antibiotics	May contain chemical residues	Organic, safe, high quality
Climate restriction level	High	Medium	Low	Extremely low (controlled indoor)
Initial investment cost	Low	Medium	High	Higher
Operating cost	Medium	Medium	Medium (water & fertilizer costs)	Lower (water, fertilizer, pesticide, labor savings)
Return on investment period	Longer	Medium	Medium	Shorter (3–5 years)

traditional aquaculture tailwater. At the same time, the system does not use fossil fuels, and its energy demand is mainly electricity, which can be supplemented by clean energy such as photovoltaics, further reducing the carbon footprint (Körner et al., 2021). This "zero-emission" characteristic is not only a manifestation of environmental responsibility but also a significant source of its economic competitiveness, as "zero emission" directly translates into "resource recovery and utilization," reducing waste treatment costs and increasing effective output.

- **No pesticides or chemical fertilizers:** The system does not rely on chemical fertilizers, pesticides, or herbicides, producing safer and healthier food, and reducing the harm of chemicals to the ecological environment and human health.
- **Biodiversity:** The greenhouse aquaponics model can form a micro-ecosystem with biodiversity, promoting ecological balance.

1.2.2.3 *Economic potential*

- **High yield and high value:** In a controlled environment, crop growth cycles are shortened, vegetable annual yield can increase by four harvests, fish farming cycles are halved, and feed conversion ratios are reduced by 20%, significantly increasing output per unit area. The combined output of high-quality fish and vegetables increases sources of economic income.
- **Reduced operating costs:** Automated operation significantly saves labor costs and reduces heavy manual labor. In the long run, the circular production model can reduce overall operating costs, especially in terms of water, fertilizer, and pesticide inputs.
- **Market advantage:** The production of organic, chemical-free, and localized products meets the growing consumer demand for healthy and sustainable food, allowing for higher market premiums.
- **Return on investment:** Although initial investment is high, through scaled production and refined management, a good return on investment can be achieved, with commercial systems typically recovering costs within 3–5 years (Love et al., 2015).
- **Employment and social benefits:** Large-scale factory projects can provide stable employment opportunities for local residents, drive regional economic development, and improve farmers' livelihoods. Despite higher initial investment costs, the economic viability of intelligent aquaponics farming lies in their ability to reduce long-term operating costs and diversify income through scale, automation, and high-value products. This requires investors to have a long-term strategic vision and an understanding of technological integration, as early high investment is recouped through later operational efficiency and market competitiveness.

1.2.3 Challenges faced and intelligent solutions

While smart aquaponics farming offers immense potential, they also face a series of challenges, and intelligence is the core strategy for overcoming these challenges.

1.2.3.1 Challenges

- **High initial investment:** Intelligent aquaponics systems require significant upfront investment, including equipment, infrastructure, and technology integration costs, which can be a barrier to widespread adoption.
- **Technical complexity and learning curve:** Maintaining the delicate balance between fish, plants, and microorganisms requires specialized knowledge and continuous monitoring. The system has numerous parameters, making management complex and requiring high technical skills from operators.
- **System fragility:** Being a closed-loop, high-density system, it is sensitive to sudden events like equipment failure or power outages, which can lead to mass mortality of fish and plants in a short period, causing significant losses.
- **Data security and technological accessibility:** The collection, transmission, and analysis of large amounts of data involve data security issues, and the cost and technical threshold of high-tech equipment and AI algorithms may hinder their widespread application (Wright et al., 2022).
- **Nutrient balance and pest/disease management:** Ensuring plants receive all necessary micronutrients and effectively controlling pests and diseases in a chemical-free environment are ongoing challenges.

1.2.3.2 Intelligent solutions

- **Lowering technical threshold and cost:** As the cost of IoT devices and AI platforms continues to decrease, and open-source solutions become more prevalent, the construction and operation of intelligent aquaponics systems are becoming more economically feasible (Wright et al., 2022).
- **Real-time monitoring and early warning:** Sensors and AI algorithms provide 24-hour uninterrupted real-time data monitoring and abnormal warnings, transforming management from reactive response to proactive intervention, significantly reducing system operational risks.
- **Automated control:** Intelligent controllers and actuators automatically adjust environmental parameters (e.g., water flow, aeration, nutrient solution dosing) based on data analysis results, reducing manual operations and human errors, and ensuring stable and precise system operation.
- **Predictive maintenance and fault diagnosis:** AI algorithms analyze equipment operation data and historical fault records to predict potential equipment failures or wear, issuing early warnings and recommending preventive maintenance, thereby avoiding production interruptions and significant losses.
- **Data-driven decision optimization:** AI models analyze data such as fish behavior and plant growth patterns to optimize feeding amounts, light intensity, and so on, improving production efficiency and resource utilization (Wright et al., 2022).
- **Remote management and traceability:** Mobile applications and web platforms enable farmers to remotely monitor and manage the system anytime, anywhere, and provide full traceability from production to table, enhancing consumer trust and product market competitiveness (Karimanzira and Jacobi, 2021).

Intelligence is the core strategy for overcoming the inherent complexity and fragility of aquaponics. Smart aquaponics farming transforms the "high initial investment" into "long-term operational efficiency and risk reduction," changing the traditional agricultural investment return model. Although initial costs are high, intelligence significantly reduces long-term operating costs and risks through automation reducing labor, precise control reducing resource waste, and predictive maintenance avoiding losses (Karimanzira and Jacobi, 2022), thereby improving return on investment. However, this still requires policy support and financial mechanisms, such as government subsidies and streamlined approval processes, to lower the initial threshold and accelerate technology adoption and commercialization.

1.3 Book structure overview

This chapter, serving as the introduction to the book, aims to provide readers with a macroscopic understanding of smart aquaponics farming, clarifying their core concepts, development history, significant advantages, and challenges faced. Through this chapter, readers will gain an initial understanding of the overall value proposition of smart aquaponics farming and their positioning in future agriculture.

Subsequent chapters will delve into the various key components of smart aquaponics farming, detailing their technical principles, intelligent management mechanisms, and performance in practical applications:

- **Chapter 2 "Digital Infrastructure System"** mainly introduces and explains the important contents of infrastructure foundation, as well as the methods and significance of realizing digital monitoring of infrastructure system.
- **Chapter 3 "Recirculating Aquaculture System"** will focus on the fish farming component, explaining the principles of recirculating water technology, intelligent water quality monitoring and regulation, and how to achieve extreme water resource recycling.
- **Chapter 4 " Hydroponic Planting System**" will explore plant cultivation techniques, including different soilless cultivation models, intelligent management of nutrient solutions, and optimization strategies for plant growth environments.
- **Chapter 5 "Air Environment Control System"** will deeply monitoring units for spatiotemporal data of greenhouse external and internal climate, as well as monitoring units for internal air environment parameters, microclimate models, and environmental parameter optimization and control technologies.
- **Chapter 6 "Fish Precision Feeding System"** will explain how to achieve precise feeding based on fish growth stages and feeding demands through intelligent equipment and algorithms, maximizing feed utilization and reducing water pollution.
- **Chapter 7 "Equipment and Energy Consumption Control System"** will introduce key production equipment within the factory and focus on how to optimize energy efficiency through intelligent means, integrate renewable energy, and perform predictive maintenance to ensure stable system operation.

- **Chapter 8 "Microbial Regulatory System"** will reveal the critical role of microbial communities in aquaponics systems. From the aspects of bacteria, fungi, viruses, and microalgae, the threats posed by pathogenic microorganisms are sorted out and described, as well as their monitoring and prevention measures.
- **Chapter 9 " Smart Aquaponics Farming Integration"** will elaborate on how intelligent aquaponics farming integration for intelligent operation through sensor networks, data acquisition, and transmission, and Big data cloud platform driven by data and AI models, achieving significant overall efficiency improvements and system optimization through intelligent integrated control.
- **Chapter 10 "Sustainability Evaluation of Smart Aquaponics Farming System"** will establish a comprehensive assessment framework to quantify the economic, net-zero, ecological and social benefits of aquaponics farming.
- **Chapter 11 "Case Studies of Aquaponics System Worldwide"** will showcase successful practices of advanced aquaponics farming through international cases, select representative aquaponics farming from countries on various continents, introduce their basic situation, and summarize the system characteristics and experience practices.

Through the systematic exposition in this book, readers will gain a comprehensive understanding of the scientific principles, technical details, application prospects, and profound impact of smart aquaponics farming on the future development of agriculture.

References

Benis, K., Ferrão, P., 2017. Potential for environment/resources-savings of urban agriculture as a form of urban ecological sanitation: a case study of Lisbon, Portugal. Journal of Cleaner Production 154, 365–375.

Delaide, B., Goddek, S., Gott, J., Soyeurt, H., Jijakli, M.H., 2016. Lettuce and basil growth performance in a decoupled aquaponic system. Horticulturae 2 (1), 2.

FAO, 2020. The State of World Fisheries and Aquaculture 2020. Sustainability in action. FAO, Rome.

Goddek, S., Delaide, B., Mankasingh, U., Ragnarsdottir, K.V., Jijakli, M.H., Thorarinsdottir, R., 2015. Aquaponics: an industrial application of closed loop systems. Systems 3 (3), 33–55.

Graber, A., Junge, R., 2009. Aquaponic systems: nutrient recycling from fish wastewater by combined fish and plant farming. Desalination 246 (1-3), 147–156.

Karimanzira, D., Jacobi, M., 2021. Electronic management of a recirculating aquaculture system. Aquacultural Engineering 92, 102143.

Karimanzira, D., Jacobi, M., 2022. Digital twin-based control and monitoring of an aquaponics system. Computers and Electronics in Agriculture 197, 106967.

Khandakar, A., Elzein, I.M., Nahiduzzaman, M., Ayari, M.A., Ashraf, A.I., Korah, L., et al., 2024. Smart aquaponics: an innovative machine learning framework for fish farming optimization. Computers and Electrical Engineering 119, 109590. Available at https://doi.org/10.1016/j.compeleceng.2024.109590.

Kyaw, T.Y., Ng, A.K., 2017. Smart aquaponics system for urban farming in Singapore. Proceedings of the IEEE International Conference on Industrial Engineering and Engineering Management (IEEM). IEEE, pp. 1141–1145.

Körner, O., Holst, N., Zwart, H.F.D., 2021. Energy saving potential of climate control in greenhouses. Agricultural Systems 190, 103061.

Li, Z., Wang, J., Higgs, R., Zhou, J., Tan, C.J., 2020. Design of an intelligent aquaponics monitoring system based on IoT and deep learning. Journal of Sensors 2020, 1–15.

Love, D.C., Fry, J.P., Genello, L., Hill, K.E., Frederick, J.A., Li, X., et al., 2015. An international survey of aquaponics practitioners. PLoS ONE 10 (7), e0133647.

Mahon, N., Raman, S., Fraile-Navarro, D., 2022. The role of artificial intelligence in sustainable agriculture: a systematic literature review. Journal of Cleaner Production 352, 131537.

Palm, H.W., Knaus, U., Appelbaum, S., Goddek, S., Strauch, S.M., Vermeulen, T., et al., 2018. Towards commercial aquaponics: a review of systems, designs, scales and components. Fisheries and Aquatic Sciences 21 (1), 1–16.

Somerville, C., Cohen, M., Pantanella, E., Stankus, A., Lovatelli, A., 2014. Small-scale aquaponic food production: iIntegrated fish and plant farming. FAO Fisheries and Aquaculture Technical Paper No. 589

Wongkiew, S., Hu, Z., Chandran, K., Kim, J.W., Khanal, S.K., 2017. Nitrogen transformations in aquaponic systems: a review. Aquacultural Engineering 76, 9–19.

Wright, S., Hughes, V., James, C., 2022. IoT and AI for Smart Aquaponics and Hydroponics Systems. 1–4.

CHAPTER

Digital infrastructure system

2

Daoliang Li[1,2] and Liang Wang[1,3]

[1]*National Innovation Center for Digital Fishery, China Agricultural University, Beijing, P.R. China,*

[2]*College of Information and Electrical Engineering, China Agricultural University, Beijing, P.R. China,*

[3]*School of Chemistry and Chemical Engineering, University of Surrey, Guildford, United Kingdom*

2.1 Foundational principles of smart aquaponics farming infrastructure

2.1.1 Strategic site selection and physical infrastructure

The selection of a site and the design of the physical infrastructure for a smart aquaponics farming must move beyond the passive mindset of traditional farming. For large, digitally controlled facilities, site selection criteria shift from passive acceptance to active design considerations.

The primary principle is geological and environmental stability (Palm et al., 2018). Given the immense weight of aquaponic systems, especially the water and growing media, it is essential to choose a stable, level surface to prevent the risk of system legs sinking, disrupting water flow, or even a catastrophic collapse. To enhance stability, concrete slabs or cement blocks can be placed under the system's supports, and crushed stone can be used to level and stabilize the ground.

In addition to geological stability, external environmental factors like wind, rain, and snow are no longer risks the farming must passively endure (FAO, 2014). Strong winds can damage plant stems, while heavy rain and snow can dilute nutrient water, flood the system, and harm unprotected electrical equipment. Therefore, a smart farming should be located in a wind-protected area and actively manage these environmental risks by building greenhouse or shade net structures. Greenhouses effectively protect against extreme weather and can extend the growing season year-round by trapping solar heat or using internal heating. In tropical regions, a more ventilated net house structure is effective for preventing issues with excessive heat and humidity.

For a smart farming, the accessibility of utilities and logistics is paramount. Beyond conventional roads and water sources, stable electricity and network connectivity are the cornerstones of the digital infrastructure. All power outlets must be equipped with a Residual-Current Device and protected from water to ensure personal safety.

Smart Aquaponic Farming. DOI: https://doi.org/10.1016/B978-0-443-28945-3.00004-9

The water source should be easily accessible and must be pre-treated to remove chlorine or chloramines, which are toxic to fish, plants, and beneficial bacteria, before entering the system. Stable and continuous network connectivity is no longer an optional convenience but a fundamental guarantee for all digital control, remote monitoring, and automated processes.

2.1.2 Core operational constraints and digital management

An aquaponics system is a complex, multidisciplinary ecosystem, and its central challenge is to maintain the delicate symbiotic balance between fish, plants, and nitrifying bacteria. Therefore, the farming's digital infrastructure must be built around the precise management of these biological and chemical constraints. The system's core constraints include the following:

- **Water quality parameters**: The pH (typically between 6.5 and 7.5), temperature (20°C–26°C), and dissolved oxygen (>5 mg/L) of the water must be strictly controlled (Somerville et al., 2014). The concentration of nutrients like ammonia, nitrites, and nitrates also requires continuous monitoring to prevent toxic buildup.
- **Fish-to-plant ratio**: The system must maintain a dynamic balance between fish biomass, plant density, and feed input. By precisely controlling these parameters, the system ensures that fish waste provides sufficient nutrients for the plants, while the plants effectively purify the water.
- **Waste and wastewater management**: Although aquaponic systems are highly water-efficient, they still require periodic water changes and filter flushing. A digital system can optimize this process, for example, by directing nutrient-rich wastewater to nearby soil-based plants for reuse.

Modern digital management systems are capable of effectively addressing these challenges. Internet of Things (IoT) sensors can continuously collect water quality data and transmit it in real time to a central control system. When parameters deviate from the normal range, the system can automatically activate actuators like pumps or valves to make immediate adjustments (Dubey, 2025). This automation and precision control are crucial for ensuring the efficient and stable operation of a large-scale farming.

2.2 The digital nervous system: integrated sensors and monitoring architecture

2.2.1 The Internet of Things framework

The digital infrastructure of a smart farming is built on an IoT framework. It comprises a network of interconnected sensors and actuators, forming a vast network capable of large-scale data collection and real-time control. The data collected by these sensors, such as fish tank nutrient concentration or battery load cycles, is sent to the cloud on a per-time-step basis for remote monitoring and analysis via a cloud-based dashboard.

The application of IoT has three evolutionary stages (Karimanzira and Jacobi, 2021), each bringing a fundamental shift in the operational paradigm. Initially, IoT was used for passive monitoring, where sensors simply collected data and displayed it on a dashboard. While this aided human observation, it still required manual intervention to interpret the data and perform actions.

The second stage is automated control, where the system directly links sensors with actuators. For example, when pH or turbidity levels exceed a threshold, a pump can automatically be activated to adjust or replenish the water. This automated response improves efficiency and reduces human error.

Ultimately, the evolution of IoT technology leads to the creation of a Digital Twin. A Digital Twin is a dynamic virtual copy of a physical system that is continuously updated with real-time sensor data and can simulate and optimize operations. This not only allows for real-time monitoring of system performance but, more importantly, shifts the operational paradigm from a passive, reactive approach to a proactive, predictive one. With a Digital Twin, operators can test different variables in a virtual environment, predict potential problems, and take preventive measures before they occur, significantly improving the farming's ecological and operational efficiency.

2.2.2 Underground and pipe network sensing

The digital infrastructure of a smart aquaponics farming faces a unique challenge: monitoring the pipeline systems that are buried underground or are difficult to access. To ensure the continuous delivery of critical fluids like water, oxygen, heat, and feed, the pipeline network must be continuously monitored.

Specialized IoT sensors are used to monitor the pressure and flow in buried pipes. These sensors often use low-power, long-range communication technologies like LoRaWAN, LoRaWAN can cover a wide area, and its ultra-low power consumption allows sensors to operate on a single battery for up to 10 years in harsh or inaccessible environments. By measuring pressure and detecting anomalies, these sensors can detect leaks or blockages early. This early warning mechanism is crucial for preventing water waste and system performance degradation, nipping potential catastrophic failures in the bud.

2.2.3 Condition-based monitoring of critical equipment

To maximize equipment uptime and reduce maintenance costs, smart factories are shifting from scheduled preventive maintenance to predictive maintenance based on real-time data. The core of this approach is the continuous monitoring of the health of critical equipment using digital tools (Mobley, 2002).

By permanently installing vibration and temperature sensors on key rotating equipment, such as water pumps, air pumps, and climate control fans, their operational status can be continuously monitored. Every machine has a characteristic, non-anomalous vibration "signature." When an anomaly in operation occurs, this vibration signature changes. By analyzing vibration data, such as evaluating metrics like root mean square acceleration and crest factor, engineers can predict when common issues like bearing wear, component misalignment, or motor failure are likely to occur before they happen.

Predictive maintenance systems eliminate unnecessary and premature parts replacements, thereby reducing maintenance costs and inventory requirements. By providing advance notice of potential equipment failures, maintenance teams can schedule repairs more effectively, intervening without disrupting production and minimizing unplanned downtime.

2.2.4 Ground-level asset identification and status detection

In a large farming, managing thousands of pieces of equipment and components is a daunting task. The digital infrastructure solves this problem by giving each ground-level asset a unique identity marker, enabling real-time tracking and status updates. This can be achieved through technologies like RFID tags or QR codes. This integrated asset management capability is essential for automated inventory, accurate location, and monitoring the status of all equipment, making it an indispensable part of high-density production operations.

To provide a more concrete illustration of predictive maintenance in practice, the following Table 2.1 links key aquaponics equipment with their common failure modes and corresponding digital monitoring metrics.

This table translates the abstract concept of predictive maintenance into actionable strategies. By connecting specific hardware (e.g., a water pump) with potential failure modes (e.g., bearing degradation) and their corresponding precise digital signals (e.g., an increase in crest factor), it provides a concrete, deployable guide for engineers and operators. It highlights how a smart system can use non-invasive technical means to transform potential mechanical problems into predictable and manageable events, thereby ensuring the continuous and stable operation of the complex aquaponic biological ecosystem.

2.3 Autonomous operations and mobility

2.3.1 Automated guided systems for farming logistics

Automated guided vehicles (AGVs) play a critical role in a smart aquaponics farming by automating material transport and crop management tasks within the

Table 2.1 Predictive maintenance metrics for aquaponics equipment.

Equipment type	Key component	Common failure mode	Vibration analysis metrics & signals	Alert trigger	Operational value
Water pumps, air pumps	Bearings	Wear, damage	Increased crest factor; abnormal increase in RMS acceleration	Monitored value exceeds baseline threshold	Prevents catastrophic failure, reduces repair costs
Water pumps, motors	Shaft	Misalignment, imbalance	Increase in amplitude of harmonics or sidebands of the running speed frequency (1 × RPM)	Vibration amplitude exceeds normal range	Improves system reliability, avoids water flow issues due to pump failure
Climate control fans	Blades	Cracks, missing	Increase in harmonics of the blade pass frequency (rotation speed x number of blades)	Appearance of abnormal frequencies	Avoids fan failure leading to loss of temperature or humidity control, protects animal and plant health
Heating/cooling systems	Insulation	Aging, damage	Continuous abnormal temperature sensor readings	Temperature reading deviates from set range	Ensures stable water temperature, prevents heat stress and disease outbreaks

facility. To optimize operational efficiency, a detailed comparison of AGV navigation technologies is necessary.

- **Magnetic tape/spot navigation**: This method is low-cost and easy to install, but it has a clear vulnerability in the wet environment of aquaponics. While short-term water exposure won't destroy the magnetic strip, it can weaken the tape's adhesion and cause it to warp from uneven drying. Fixed paths also lack flexibility.
- **QR code navigation**: QR code navigation offers high flexibility and adaptability. It is easy to set up and modify, but its reliability is highly dependent on a clean, well-lit environment, which can be challenging in a wet or inconsistently lit farming.
- **Laser-guided navigation (LGV)**: LGV is a high-precision, highly reliable navigation method. It uses a laser scanner on the AGV to triangulate its position with fixed reflectors in the environment. This system does not rely on floor markers and is non-invasive to the floor, making it well-suited for complex farming layouts.
- **Natural navigation (SLAM)**: This is the most flexible method, where the AGV builds a map of the environment in real time with its sensors and self-localizes. It eliminates the dependence on any fixed paths and is considered the core technology for automated mobile robots (AMRs).

There is an interesting trade-off between efficiency and flexibility between AGVs and AMRs. Although AMRs gain greater flexibility with SLAM technology and can freely avoid obstacles, this "unrestricted" movement can be a weakness in a structured, repetitive farming environment. AGVs that follow a predetermined path can achieve predictable, high-speed, and high-throughput operations. In contrast, a fleet of unmanaged, free-roaming AMRs in a high-density environment could lead to traffic chaos, reducing overall efficiency. Therefore, for an aquaponics farming, the optimal solution may not be to simply choose the "most advanced" technology, but a hybrid approach that combines laser navigation or natural navigation with centralized traffic management software (Bechar and Vigneault, 2016). This method can achieve both efficient logistics and the ability to respond to unexpected situations in a dynamic environment.

2.4 Automated disaster recovery and business continuity

2.4.1 Backup power systems

To protect the farming's biological assets during a power outage, a multi-layered backup power strategy must be established. The core of this is the automatic transfer switch (ATS) (Kiran et al., 2023). An ATS is an intelligent, self-acting power switching device that continuously monitors the main power source's voltage and frequency. When the primary power is interrupted, the ATS will immediately trigger the backup generator to start and will automatically switch the load to the generator within seconds after it stabilizes. When the main power is restored, the ATS automatically switches the load back and allows the generator to enter a cool-down period before shutting down.

A more advanced system goes beyond simple switching and is managed by a Programmable Logic Controller (PLC). It can execute a load shedding protocol. This means that when the generator capacity is limited, the system can connect loads in a pre-programmed priority sequence, ensuring that the most critical systems, such as oxygen and water pumps, receive power first. Less critical loads, like lighting and Heating, Ventilation and Air Conditioning (HVAC), might be automatically disconnected to ensure the survival of the core biological system. This intelligent, tiered disaster recovery strategy elevates a simple power outage prevention system into a life support system that can handle a variety of situations and guarantee the survival of the core ecosystem.

2.4.2 Water and oxygen management

Beyond backup power, water and oxygen management are equally important components of disaster recovery. For fish, a lack of oxygen is lethal, so a continuous oxygen supply is critical. A smart farming can achieve complete oxygen autonomy using on-site oxygen generation systems, which eliminates reliance on external liquid oxygen deliveries that can be risky due to logistical disruptions.

In an emergency, in addition to a backup oxygen generation system, battery-powered air pumps can be used for short-term oxygenation. In addition, high-capacity oxygen diffusers can be automatically activated when sensors detect a drop in dissolved oxygen levels, quickly injecting pure oxygen stored in tanks into the water. This buys valuable time for operators to address the root problem. For water management, an automated system can use ultrasonic sensors to monitor water levels. If a drop-in water level is detected, the system can automatically start pumps to replenish it. Similarly, when water quality parameters like pH, temperature, or dissolved oxygen deviate from the optimal range, actuators will automatically make corrections.

2.5 Human–machine interface: augmented reality for operations and maintenance

2.5.1 Real-time data visualization and interaction via augmented reality

While a remote monitoring dashboard provides convenience, the future of smart factories lies in a more intuitive, immersive human-machine interface. Augmented reality (AR) technology achieves this leap by overlaying a virtual data layer onto physical objects in the real world. This provides operators with a new way of interacting: Instead of switching between multiple screens and complex charts, they can simply point their phone or AR glasses at a piece of equipment and instantly see real-time data, such as water quality metrics, temperature readings, or equipment status (Huuskonen and Oksanen, 2018).

This immersive data overlay is not limited to visual information. The AR system can also use AI algorithms to provide precise fertilization and watering decisions for plants, using data from plant sensors, weather forecasts, and soil moisture levels. Furthermore, AR glasses offer a hands-free way of working, which is particularly crucial when performing maintenance and inspection tasks.

2.5.2 Practical applications of augmented reality in smart aquaponics farming

The application of AR in a smart aquaponics farming extends beyond simple data display; it is fundamentally changing the way operations and maintenance are conducted.

- **Enhanced inspections**: AR can be used for "crop scouting," where an operator can wear AR glasses or use a mobile device to simply point their camera at a plant to see a real-time data overlay of its health status, nutrient levels, and potential diseases. This data is automatically identified by sensors or computer vision analysis. This method greatly simplifies plant health assessment, making early and targeted intervention possible.
- **Remote assistance and training**: AR glasses allow on-site technicians to receive step-by-step guidance from remote experts via a live video stream while performing equipment maintenance. This not only reduces the need for experts to travel, lowering costs, but also significantly speeds up repair times. It is also a powerful tool for training new employees, providing a more engaging and effective learning experience than traditional manuals.

The core value of AR is its ability to bridge the long-standing "knowledge gap" in the aquaponics field. An aquaponics system is a complex ecosystem that requires operators to possess a combination of knowledge from biology, botany, plumbing, and electrical engineering. This is a major challenge for commercial adoption due to the lack of experienced labor. AR technology effectively "democratizes" this expertise by transforming complex real-time data into a simple, intuitive, and actionable format. It allows less experienced operators to perform complex tasks by providing hands-free, guided workflows and instant access to expert knowledge, directly addressing the labor shortage and improving efficiency while lowering operational costs.

2.6 Conclusion and strategic recommendations

2.6.1 The integration of a fully integrated smart aquaponics system

The digital infrastructure of a smart aquaponics farming is an interconnected ecosystem of several complex subsystems (Muhammad et al., 2024). At its core is a powerful IoT framework that acts as a digital nervous system (Ibrahim et al., 2023), continuously collecting real-time data through a network of sensors spread throughout the farming. This data is transmitted to a central control system, such as a PLC, which acts as the brain, analyzing the data and coordinating the farming's "body"—the AGVs, disaster recovery systems, pumps, and valves (Risma et al., 2025). At the same time, operators interact with the entire system through an augmented reality interface. This interface goes beyond a traditional data dashboard, directly overlaying real-time information, instructions, and expert guidance onto the physical world, creating an intuitive and immersive control experience. This comprehensive integration transforms the aquaponics farming into a highly efficient, resilient, and automated production entity.

2.6.2 Future outlook and scalability considerations

The digital infrastructure of a smart aquaponics farming is key for modern agriculture to address global challenges. By precisely controlling resources, maximizing productivity, and reducing reliance on labor and external supplies, it contributes to food security and sustainable development. These technologies make it possible to achieve efficient and scalable food production in non-traditional areas like cities.

From a strategic perspective, a phased implementation approach is recommended. First, establish a robust IoT foundation to enable basic data monitoring and automated control. Subsequently, gradually integrate more complex systems like predictive maintenance and automated guided vehicles to optimize operational efficiency. Finally, invest in cutting-edge technologies like augmented reality and digital twins to achieve the highest level of operational intelligence and human–machine interaction. This phased strategy can help an organization manage its initial investment while progressively moving toward an efficient, resilient, and intelligent aquaponics production system that is capable of adapting to future needs.

References

Bechar, A., Vigneault, C., 2016. Agricultural robots for field operations: concepts and components. Biosystems Engineering 149, 94–111.

Dubey, A., 2025. aquaponics. Encyclopedia Britannica. Available at https://www.britannica.com/technology/aquaponics.

FAO, 2014. *Small-Scale Aquaponic Food Production: Integrated Fish and Plant Farming.* FAO Fisheries and Aquaculture Technical Paper No. 589. Food and Agriculture Organization of the United Nations.

Huuskonen, J., Oksanen, T., 2018. Augmented reality for supervising autonomous machines in smart farming. Computers and Electronics in Agriculture 153, 353–359.

Ibrahim, W., Abdul Rasid, R., Hariyanto, N., Faizal, N., Mazlan, N., 2023. Smart home aquaponic system monitoring and control with Internet of Things using mobile application. Journal of Chemical Engineering and Industrial Biotechnology 9, 20–25.

Karimanzira, D., Jacobi, M., 2021. Electronic management of a recirculating aquaculture system. Aquacultural Engineering 92, 102143.

Kiran, N.U., Naik, M.G., Aitwar, V.S., Madhu, K., Kumar, A.V., 2023. Smart Aquaponic System: Usage of Artificial Intelligence in Aquaculture. Trends In Agriculture Science 2 (7), 533–535. Available at https://doi.org/10.5281/zenodo.8152495.

Mobley, R.K., 2002. An Introduction to Predictive Maintenance. Butterworth-Heinemann.

Muhammad Ezaan Ali, J., Sultan, K., Hafeez, A., 2024. A smart monitoring and observation framework for aquaponics ecosystem. .ARPN Journal of Engineering and Applied Sciences 19 (13), 864–869.

Palm, H.W., Knaus, U., Appelbaum, S., Goddek, S., Strauch, S.M., Vermeulen, D., Kotzen, B., 2018. Towards commercial aquaponics: a review of systems, designs, scales and components. Fisheries and Aquatic Science 21 (1), 1–16.

Risma, P., Prasetyo, T., Utami, P., Sianipar, A., Rahman, R., Hibrizi, D., 2025. Smart fish feeding solutions for aquaponic farming: load cell and real-time clock integrated system design. Proceedings of the 8th FIRST 2024 International Conference on Global Innovations (FIRST-ESCSI 2024). Atlantis Press, pp. 91–113. Available at https://doi.org/10.2991/978-94-6463-678-9_11.

Somerville, C., Cohen, M., Pantanella, E., Stankus, A., Lovatelli, A., 2014. Small-scale aquaponic food production: integrated fish and plant farming. FAO Fish. Aquaculture Technical Pap. No 589.

CHAPTER

Recirculating aquaculture system

3

Daoliang Li[1,2], Cong Wang[1], Lingwei Jiang[1,3] and Chang Liu[1]

[1]*National Innovation Center for Digital Fishery, China Agricultural University, Beijing, P.R. China,*
[2]*College of Information and Electrical Engineering, China Agricultural University, Beijing, P.R. China,*
[3]*School of Chemistry and Chemical Engineering, University of Surrey, Guildford, United Kingdom*

3.1 Recirculating aquaculture technology

Aquaponics integrates aquaculture with crop cultivation through engineered designs, establishing a symbiotic cycle between plants, microbes, and fish. The aim is to combine fish-rearing and vegetable-growing environments within a single system. Aquaponic systems are commonly categorised as either coupled aquaponic systems (CAS) or decoupled aquaponic systems (DAS). The principle of coupled aquaponics was first described by Huy Tran at the World Aquaculture Conference in 2015 (Tran, 2015). The basic configuration of a coupled aquaponic system consists of one or more fish tanks, a sedimentation or clarification unit, substrates for microbial growth or biofilters, and a hydroponic unit for plant cultivation. These components are interconnected by a piping network, forming a closed loop water cycle. Typically, after mechanical filtration and biofiltration, water is collected in a sump, the lowest point of the system, where it is pumped back to the fish tanks. From there, it flows naturally by gravity into the hydroponic unit (Goddek et al., 2019). Although the structural elements of decoupled aquaponic systems are similar to those of coupled systems, the direction of water flow and the circulation method differ. In DAS, the aquaculture and hydroponic units operate with independent water and nutrient cycles, which allows more precise regulation of water chemistry within each subsystem.

Among aquaponic systems, the coupled aquaponics system represents the archetypal form. It requires advanced water treatment processes such as filtration, ultraviolet disinfection for microbial control, automated feeding, computerized monitoring, and strict biosecurity measures. Under conditions of fully closed recirculation with nutrient-enriched water, the symbiotic community of fish, plants, and bacteria can achieve higher yields than standalone fish production or plant cultivation (Goddek and Keesman, 2018). In contrast, the decoupled aquaponics system is based on independent recirculating loops for fish, plants, and biofilters (aerobic or anaerobic), connected through a unidirectional flow of water and nutrients. This design improves the recovery and bioavailability of macro- and

Smart Aquaponic Farming. DOI: https://doi.org/10.1016/B978-0-443-28945-3.00010-4

micronutrients while optimising water consumption. Decoupled aquaponics has emerged as a major research focus and development trend due to its operational flexibility. Each unit can function independently, and in cases of pathogen outbreaks or production problems, subsystems can be isolated to enhance overall control and efficiency while reducing cross-interference.

Advanced water treatment technologies play a pivotal role in aquaponics by sustaining environments that support both aquatic and plant growth. These technologies optimize the recycling of water and nutrients, thereby improving system efficiency and reducing environmental impacts. The water circulation process can be further enhanced through the application of Internet of Things (IoT) technology. Using multidimensional integration circuits (MD-IC), IoT systems enable compact, real-time monitoring of key parameters such as temperature, conductivity, ammonia, and pH. These lightweight devices are easily deployed, reduce reliance on specialist operators, and allow farmers to maintain optimal conditions autonomously.

Biochemical treatments—such as aerobic and anaerobic microbial processes, nitrification, and denitrification—are central to aquaponics. They reduce organic loads, mineralise nutrients, and render them more accessible for plant uptake, while also offering the potential to generate biogas, mainly methane, as a renewable energy source. Nitrification and denitrification are particularly vital, as they convert ammonia excreted by fish into nitrites and subsequently nitrates, which are less toxic and directly usable by plants.

To further improve water quality, aquaponic systems may integrate additional chemical, physical, and biological treatments. Chemical methods, including ozone and hydrogen peroxide, exploit strong oxidising properties to disinfect water and system components. While sodium hypochlorite is cost-effective, it is insufficient as a sole measure for pathogen control. Physical and biological treatments such as filtration and ultraviolet radiation are also employed. UV treatment is particularly effective, as it disrupts microbial DNA, thereby safeguarding system biosecurity.

The integration of these technologies provides a comprehensive strategy for water quality management and nutrient recycling in aquaponics. This holistic approach enhances system sustainability, minimizes environmental impact, and reinforces the role of aquaponics in advancing modern agriculture.

3.2 Dissolved oxygen control system

Controlling dissolved oxygen (DO) concentrations is critical for the health and growth of aquatic organisms and plants in modern recirculating aquaculture systems (RAS) and aquaponics. DO represents the amount of oxygen dissolved in water, which is indispensable for fish respiration and the metabolic processes of beneficial microorganisms. Inadequate DO levels can stress fish, leading to reduced growth, weakened immunity, and in severe cases, mortality. Moreover, DO strongly influences the performance of nitrifying bacteria in biofilters, which convert toxic ammonia into less harmful nitrites and nitrates. Maintaining precise DO levels is therefore essential to sustain fish health and to ensure the stability and efficiency of the system as a whole.

Oxygen levels in aquaponics are commonly managed either by aeration or through the direct introduction of pure oxygen. Aeration achieves oxygen saturation at atmospheric levels, whereas pure oxygen injection enables supersaturation, which is vital for systems with high stocking densities. Supersaturation is typically achieved using gas-to-liquid oxygenators that efficiently transfer oxygen into the water, thereby ensuring adequate oxygen delivery throughout the system, from fish tanks to plant-growing units.

Two widely used technologies are low-head oxygenators and diffused aerators. Low-head oxygenators are well-suited to increasing oxygen concentrations across large water volumes, making them ideal for large-scale operations requiring consistent distribution. Diffused aerators, by generating fine bubbles, maximize the contact surface between air and water, thereby enhancing oxygen dissolution. They are particularly effective in shallower water environments.

The deployment of DO sensors is indispensable for accurate, real-time monitoring. These sensors are strategically installed at varying depths to capture spatial variations in oxygen distribution. Integrated with automated control systems, they enable dynamic adjustment of oxygen supply in response to sensor readings, ensuring stable and optimal oxygen conditions across the system.

In advanced aquaponic facilities, data management software is increasingly utilised to continuously log and analyse water quality parameters. This software not only supports real-time monitoring but also archives historical records, providing valuable insights for future optimization. Long-term data analysis helps operators identify fluctuations in oxygen demand and refine management strategies accordingly, thereby improving efficiency and productivity. Effective DO management is thus central to optimizing the health and growth environment in aquaponics, maximizing both productivity and long-term sustainability.

3.3 Three-nitrogen control system

Aquaponics is a sustainable food production system that combines aquaculture and hydroponics. In this system, the excretions from aquatic animals, such as fish, are transformed by microorganisms into nutrients that plants can absorb, thus achieving resource recycling and minimizing environmental impact. Nitrogen, an essential element for life, plays a critical role in the transformation within aquaponics systems. The nitrogen cycle, as a core biochemical process, is crucial for maintaining the system's stability and production efficiency.

3.3.1 Nitrogen transformation in aquaponics

The main source of nitrogen in an aquaponics system is feed, which is digested by the fish, and most of it is eliminated from the body in the form of ammonia nitrogen

(90%), while at the same time, ammonia nitrogen is a source of nitrogen absorbed by the plant. Total ammonia nitrogen in the water exists in the form of free ammonia (NH_3) and ammonium ionic form (NH_4^+), respectively, and there is an equilibrium between the two NH_3+$H_2O$$NH_4^+$+$OH^-$. The environmental factors affecting the dynamic equilibrium of NH_3 and NH_4^+ are mainly the temperature and pH of the water (Anthonisen et al., 1976). At pH < 7, ammonia in water exists almost exclusively in the form of NH_4^+, and at pH > 11, it exists almost exclusively in the form of NH_3, and the proportion of NH_3 increases at higher temperatures. Ammonia nitrogen can be oxidised to nitrite (NO_2^-) by nitrification in the presence of ammonia-oxidizing bacteria (AOB) (e.g., Nitrosomonas, Nitrosococcus, Nitrosospira, Nitrosolobus, Nitrosovibrio sp., etc.), and the resulting NO_2^- is oxidised to (NO_3^-) by nitrite-oxidizing bacteria (NOB) (e.g., Nitrobacter, Nitrococcus, Nitrospira, Nitrospina sp., etc.), which is then taken up and used by the plant (Ebeling et al., 2006; Panuvatvanich et al., 2009). There are two main ways in which fish produce ammonia nitrogen: One is the decomposition of organic matter, such as fish excreta and urine, into ammonia nitrogen by microorganisms through ammonification, and the other is the emission of ammonia nitrogen into the water by fish through gills. Previous studies have reported that ammonia and ammonium originating from the gills comprise 60%–90% of the total nitrogen excreted by fish (Forster and Goldstein, 1969; Rychly, 1980). Urea is also largely expelled through the gills and accounts for 9%–27% of the soluble nitrogen excreted (Clark et al., 1985).

When NH_3 accumulates to a certain level, it can cause acute toxicity to fish. To ensure the normal vital activities of fish, several studies have documented that the toxic effects of ammonia are immediate, linear, and dose-dependent (Lin and Chen, 2001; Lemarié et al., 2004; Park et al., 2007). In fish, such as tilapia, exposure to ammonia not only impacts the immune response but also leads to adverse effects such as reduced growth, liver tissue degeneration, gill hyperplasia, hyperexcitability, and, ultimately, fish mortality (Yilmaz, 2019). Similarly, in shrimp, elevated levels of ammonia in the water negatively affect nutrient assimilation, growth, immune response, oxygen transportation and consumption, moulting, and overall survival (Cheng et al., 2013). The concentration of ammonia nitrogen must be maintained below 1 mg L^{-1}; nitrite is a toxic species of inorganic nitrogen at levels more than 1 mg L^{-1}. Haywood (1983) has studied that acute toxicity can occur at 0.2 mg L^{-1} NH_3 in salmonids, and a maximum permissible level of just 0.002 mg L^{-1} NH_3 has been recommended (Haywood, 1983). Nitrate is considered significantly less toxic than ammonia and nitrite, primarily because its concentration is usually much higher than that of ammonia and nitrite, as it is the final product of nitrification. Previous studies have reported that nitrate is harmful for many freshwater fish species at concentrations only above 1000 mg L^{-1} and for marine species at concentrations more than 500 mg L^{-1} (Colt, 2006). The sum of the two (NH_4^+ + NH_3) is usually referred to as total ammonia nitrogen (TAN) or simply ammonia. It is common in aquatic chemistry to express inorganic nitrogen compounds in terms of the nitrogen they contain, that is, NH_4^+–N (ionized ammonia nitrogen), NH_3–N (un-ionized ammonia nitrogen), NO_2^-–N (nitrite-nitrogen), and NO_3^-–N (nitrate-nitrogen). This allows for easier computation of total ammonia nitrogen (TAN = NH_4^+–N + NH_3–N) and mass balances between the various stages of nitrification.

There are three pathways for the removal of ammonia nitrogen in aquaculture systems photoautotrophic algae, autotrophic bacterial conversion from ammonia nitrogen to nitrite, and heterotrophic bacterial conversion from ammonia nitrogen directly to microbial biomass. Algae can utilize ammonium nitrogen or nitrite in water through photosynthesis to convert it into their biomass and oxygen. Autotrophic bacteria, on the other hand, can transform ammonium nitrogen into nitrite through nitrification, consuming dissolved oxygen and alkalinity in the process. Heterotrophic bacteria, during the reaction, require an additional carbon source to convert ammonium nitrogen into their biomass. To improve the nitrogen utilization efficiency (NUE), the large fixed-cell biofilters are routinely used in aquaponic systems that rely on the nitrification of ammonia nitrogen to nitrate-nitrogen by AOB and NOB. Simultaneously, to achieve maximum nitrite production for plant absorption and utilization while reducing the growth of algae and heterotrophic bacteria.

Photoautotrophic (algal-based systems)

$$16NH_4^+ + 92CO_2 + 92H_2O + 14HCO_3^- + HPO_4^{2-} \rightarrow C_{106}H_{263}O_{110}N_{16}P + 106O_2 \quad (3.1)$$

$$16NO_3^- + 124CO_2 + 140H_2O + HPO_4^{2-} \rightarrow C_{106}H_{263}O_{110}N_{16}P + 138O_2 + 18HCO_3^- \quad (3.2)$$

Autotrophic bacteria (AOB and NOB)

$$NH_4^+ + 1.83O_2 + 1.97HCO_3^- \rightarrow 0.0244C_5H_7O_2N + 0.976NO_3^- + 2.9H_2O + 1.86CO_2 \quad (3.3)$$

Heterotrophic bacteria

$$NH_4^+ + 1.18C_6H_{12}O_6 + HCO_3^- + 2.06O_2 \rightarrow C_5H_7O_2N + 6.06H_2O + 3.07CO_2 \quad (3.4)$$

Two phylogenetically distinct groups of bacteria collectively perform nitrification. These two groups of bacteria are generally categorized as chemosynthetic autotrophic bacteria because they derive their energy from inorganic compounds, as opposed to heterotrophic bacteria that derive energy from organic compounds. AOB obtain their energy by catabolizing un-ionised ammonia to nitrite and include bacteria of the genera Nitrosomonas, Nitrosococcus, Nitrosospira, Nitrosolobus, and Nitrosovibrio. NOB oxidize nitrite to nitrate and include bacteria of the genera Nitrobacter, Nitrococcus, Nitrospira, and Nitrospina. Nitrifying bacteria are primarily obligate autotrophs, which consume carbon dioxide as their primary carbon source, and obligate aerobes, which require oxygen to grow. All members of the Nitrobacteraceae family are gram-negative chemoautotrophs, also known as lithotrophs. In contrast to algae, chemoautotrophic bacteria are involved in aerobic reaction processes, whereas algae produce oxygen through the reaction process.

3.3.2 Nitrogen prediction and control in aquaponics systems

The modeling methods for complex processes mainly include mathematical modelling, data-driven modeling, and hybrid modeling.

3.3.2.1 *Mathematical model for nitrogen prediction*

A mathematical model is an approach that involves constructing models based on the internal physical, chemical, or mathematical principles of a system. It requires a deep understanding of the structure and interactions of the system in order to formulate algebraic or differential equations that represent the underlying processes. However, the process of mathematical modeling often involves unknown model parameters that are difficult to obtain. These parameters can be estimated using methods such as regression or fitting. This approach is effective for describing specific reaction processes and provides robust interpretability and reliability, which can aid subsequent control and optimiation efforts. However, in cases where mechanistic knowledge of a process is difficult to obtain or accurately articulate, the construction of mechanistic models can be challenging, limiting their applicability. Li et al. established a dynamic model based on mass balance equations to explain the dynamic changes in food nutrients in a semi-intensive pond. This model encompasses food nutrients, fish growth, basic nutrients, and dissolved oxygen. Among these, basic nutrients primarily simulate the concentrations of inorganic phosphorus and inorganic nitrogen. To validate the accuracy of the model, sensitivity analysis and assessment of goodness-of-fit were employed. The simulation results showed strong agreement and indicated an increase with the addition of inorganic fertilizers (Li and Yakupitiyage, 2003).

3.3.2.2 *Data-driven model for nitrogen prediction*

In situations where the internal mechanisms are unknown or difficult to describe accurately, a mathematical model may not be appropriate for parameter prediction. However, if the external input–output data of the process are available, data-driven modelling can be used. This approach relies heavily on a significant amount of experimental or observational data, where a model is constructed by minimising the discrepancy between the predicted outputs of the model and the actual measurements of the process inputs and outputs. The distinctive feature of this model lies in its reliance on data patterns rather than underlying mechanisms. It offers broad applicability and relative simplicity, but may lack interpretability, require extensive data and have limited generalisation capabilities. Currently, commonly used data-driven models include linear regression, support vector machines (SVM), random forests, neural networks, and long short-term memory (LSTM) networks

(Li et al., 2022b). Li et al. utilized deep learning methods to compare propagation neural network (BPNN), radial basis function neural network (RBFNN), SVM, and least squares SVM (LSSVM) for simulating and predicting various water quality parameters, including dissolved oxygen (DO), pH, ammonium nitrogen (NH_3–N), nitrogen (NO_3–N), and nitrite nitrogen (NO_2–N). The results revealed that the SVM model exhibited the most favorable predictive performance across all four parameters, achieving an accuracy rate of 99% for each (Li et al., 2022a). Markus et al. employed Artificial Neural Networks (ANN), Evolutionary Polynomial Regression (EPR), and Naive Bayes Model (NBM) to predict nitrite concentrations in the agricultural watershed of central Illinois. The results indicated that, for specific forecasting purposes, the ANN model with two hidden nodes (ANN2) exhibited the highest accuracy in terms of Root Mean Square Error (RMSE) (Markus et al., 2010). Kamali et al. proposed integrating moving horizon estimation (MHE) with nonlinear model predictive control (NMPC) for an integrated closed-loop approach. They conducted simulations to assess the control performance under two fault scenarios: aeration system failure in a moving bed biofilm reactor (MBBR) and oxygen sensor failure in the fish tank. The results demonstrated that in the presence of these faults, utilizing the MHE–NMPC controller effectively maintained the concentrations of ammonia nitrogen and nitrogen within acceptable levels (Kamali et al., 2023).

3.3.2.3 Hybrid model for nitrogen prediction

A hybrid model combines the strengths of both mechanistic and data-driven models to achieve more accurate predictions. Utilizing a mechanistic model as the foundation enhances the model's generalization ability. For instance, the chemical reaction rate parameters from the mechanistic model can be integrated with statistical relationships from the data-driven model, resulting in improved predictions of nitrogen transformation. The advantage of hybrid models lies in their amalgamation of physical mechanisms and data fitting capabilities. However, it requires careful management of the balance between the two and addressing the complexity that arises from integrating different modeling approaches. Liu et al. proposed a multi-fidelity surrogate modeling method based on a physics-informed neural network (PINN), which was compared with a feedforward neural network (FNN) model. The results indicate that, for data sizes of 200 and 300, the predictive performance of the two models is comparable when the data size is 300. However, with a data size of 200, the MFPINN model demonstrates a superior predictive performance (Liu et al., 2023).

3.3.2.4 Control of nitrogen dynamics in aquaculture

To enhance the intelligent and precise management of fish-vegetable symbiotic systems, adopting control strategies can effectively improve system stability, ensuring fish well-being, optimal growth conditions, and economic benefits. Applying model predictive control to the dynamic management of nitrogen can help achieve intelligent regulation of nitrogen. Research has been conducted to regulate nitrogen concentrations in water using diverse control methodologies. Lee et al. employed a control system based on fuzzy logic to optimize denitrification rates and effectively mitigate the discharge of toxic by-products, namely nitric oxide, nitrous oxide, nitrogen dioxide, or hydrogen sulfide. This fuzzy logic-based control system adeptly maintained low concentrations while preventing an increase in nitric dioxide and hydrogen sulfide levels (Lee et al., 2000). In a study by Almeida et al. combination of PID control, linearising control, and cascade control strategies (integrating PI and linearising controllers, with the PI controller implemented in the outer loop) was applied to regulate levels in the denitrification reactor of a RAS through the controlled addition of acetic acid. The study's findings highlighted that the introduction of measurement noise adversely affected PID performance to a greater extent compared to its impact on a linearising controller (Almeida et al., 2020).

Due to the integrated nature of water recirculation in aquaponic systems, these systems exhibit nonlinearity and high interactivity. Furthermore, traditional control methods such as PID control, PI control, or fuzzy control fail to adequately consider water quality constraints, which are crucial for the well-being of fish. Therefore, there is a need for advanced model-based multi-input multi-output (MIMO) controllers. In the process of data acquisition, challenges arise from the limited precision and high costs associated with certain measurement devices, hindering comprehensive system access (Fowler et al., 1994). To address this issue, dynamic model-based state estimation techniques can be employed to estimate unobservable states. These techniques utilize both measured system values and a mathematical model that relates the measured and unmeasured variables to estimate the unobservable states (Alexander et al., 2020). Several established estimation schemes exist, including Kalman filtering (KF), extended Kalman filtering (EKF), unscented Kalman filtering (UKF), ensemble Kalman filtering (EnKF), particle filtering (PF), and moving horizon estimation (MHE) (Alexander et al., 2020). Among these estimators, moving horizon estimation (MHE) stands out as an appealing choice for RAS due to its capability to accommodate the intricate behaviour of RAS and account for biological constraints within the system. Considering its potential, it is worthwhile to explore its application in the context of aquaponic systems.

3.4 pH control system

pH is a measure of how acidic/basic water is. The range goes from 0 to 14, with 7 being neutral. pHs of less than 7 indicate acidity, whereas a pH of greater than 7 indicates a base. pH is really a measure of the relative amount of free hydrogen and hydroxyl ions in the water. Water that has more free hydrogen ions is acidic, whereas water that has more free hydroxyl ions is basic. As pH can be affected by chemicals in the water, pH is an important indicator of water that is changing chemically. pH is reported in "logarithmic units." Each number represents a 10-fold change in the acidity/basicness of the water. Water with a pH of 5is ten times more acidic than water having a pH of 6. The pH of water determines the solubility and biological availability of chemical constituents such as nutrients (phosphorus, nitrogen, and carbon) and heavy metals (lead, copper, cadmium, etc.). For example, in addition to affecting how much and what form of phosphorus is most abundant in the water, pH also determines whether aquatic life can use it. In the case of heavy metals, the degree to which they are soluble determines their toxicity. Metals tend to be more toxic at lower pH because they are more soluble.

Excessively high and low pHs can be detrimental to the use of water. High pH causes a bitter taste, water pipes and water-using appliances become encrusted with deposits, and it depresses the effectiveness of the disinfection of chlorine, thereby causing the need for additional chlorine when the pH is high. Low-pH water will corrode or dissolve metals and other substances. Pollution can change a water's pH, which in turn can harm animals and plants living in the water. For instance, water coming out of an abandoned coal mine can have a pH of 2, which is very acidic and would definitely affect any fish crazy enough to try to live in it! By using the logarithm scale, this mine-drainage water would be 100,000 times more acidic than neutral water—so stay out of abandoned mines.

3.5 Temperature control system

Each aquaculture farm has a unique optimum temperature range. Fish can be classified into three classifications depending on their temperature preference: cold-water, cool-water and warm-water species. For cold-water species, optimum water temperatures are below 15°C, for cool-water species between 15°C and 20°C, and for warm-water species above 20°C. Temperature affects the physiological processes of the reared species, such as respiration rate, efficiency of feeding, growth, behavior, reproduction, and depuration. In addition, several factors affect the tolerance of fish to different temperatures, including species, age, size, and past thermal history. In RAS, temperature should be measured from several locations, such as in the rearing tank water and the slaughtered fish. Especially, measurement of inlet water temperature is important for designing and selecting a suitable heating equipment and for operating an RAS. The control of rearing temperature allows maintaining good rearing conditions for optimum growth and feed utilization, although the inlet water temperature may vary considerably between seasons. A monitoring equipment based on a programmable logic controller (PLC) for real-time monitoring of recirculating water and water level was reported by Wu et al. (2010). On the other hand, water source heat pumps generate low-grade heat from natural sources and can be used for heating and cooling. For heating, four different heating methods are typically used; boilers, geothermal energy, factory waste heat and heat pumps or other equipment. Industrial heat is limited by geographical conditions. Heating by boilers is common, but increasing attention to environmental pollution is likely to reduce the attractiveness of boiler heating.

So far, only a limited number of studies have been related to temperature control in RAS. Water source heat pumps have been used to heat aquaculture water and to maintain a stable temperature around the year, while Good et al. (2011) used a heat exchanger in RAS rearing rainbow trout (*Oncorhynchus mykiss*) and Davidson et al. (2014) a geothermal heat exchanger. Geothermal energy has been used for heating in RAS farms producing salmon, trout, tropical fish, lobsters, shrimp, and prawns. Waste heat from industry has been used for commercial oysters, penaeid shrimp, and salmon farming, while heat from power plants has been used for eel and salmonid fingerlings production. In addition, ground heat has been used as a heat source or temperature control in RAS. In recent years, there has been interest in solar technologies to maintain selected water temperature in RAS, but only a simulation model regarding solar heating systems in RAS has so far been published.

3.6 Water level control system

Water testing is essential to confirm and maintain good water quality in a system. Bacteria cannot be seen or measured directly. Therefore, water testing is the only method of indirectly diagnosing bacteria's health and activity. A grower should design and follow a schedule for water testing. A new grower is encouraged to check water quality parameters daily. During the initial system startup or the first 2 months, ammonia and nitrite should also be tested daily and then weekly thereafter to monitor the health of bacteria and fish populations. Ammonia and nitrite should also be tested anytime abnormal fish mortality occurs to rule out potential toxicity due to a bacterial population collapse. Once the nutrients have levelled off and the system is considered in balance, measurements can be taken weekly. The first step when planning a new aquaponic operation is to determine the quality of available water. If poor water is the only available source, aquaponics will never be profitable. Therefore, it is important to know what your water quality—or more precisely, water chemistry—is first, before you expose fish or plants to it. There are four possible sources of water for use in an aquaponic system: well water, municipal water, rainwater, and reverse osmosis (RO) water. Surface water is not recommended, as it may contain high levels of solids, unwanted fish or insect larvae, aquatic vegetation, including algae, and possibly pathogenic microorganisms.

A microprocessor-controlled frequency state change detection method uses a low-amplitude signal generated on multiple sensor probes of differing lengths. Each probe has a frequency separate from all other probes in the array and independently changes state when touched by water. The state change of the frequency on each probe is monitored by a microprocessor, which can perform multiple water-level control functions. A strength of state-dependent frequency monitoring is the long-term stability of the sensing probes. The signal strength is not sufficient to cause fouling, degradation, or deterioration of the sensors due to electrolysis in contaminated water. Sensor cleaning requirements are minimal or eliminated. The use of multiple sensing rods of different lengths allows the user to intuitively set up control switches at various water heights. Researchers use an electronic sensor that contains multiple stainless steel probes that are inserted into the water-based liquid being measured. These sensor probes never foul, plate, or degrade, no matter the water quality. The electronic sensor array is connected to our controller, which runs software to determine the water level and activate relays accordingly. The relays can be connected to any operation, but are primarily used for valves, pumps, and alarms. Each probe on the sensor array is responsible for a different function. For example, the WLC 6000 sensor array contains six individual probes that cut out the water heater, activate a "low alarm," turn on a fill valve, turn off a fill valve, and activate a "high alarm." There is also a complete line of pump-down controls. Our various sensors contain different combinations of these functions, and many are custom designed to fit the needs of the customer's requirements.

References

Alexander, R., Campani, G., Dinh, S., Lima, F.V., 2020. Challenges and opportunities on nonlinear state estimation of chemical and biochemical processes. Processes 8 (11), 1462. Available at https://doi.org/10.3390/pr8111462.

Almeida, P., Dewasme, L., Wouwer, A.V., 2020. Denitrification control in a recirculating aquaculture system—a simulation study. Processes 8 (10), 1–23. Available at https://doi.org/10.3390/pr8101306.

Anthonisen, A.C., Loehr, R.C., Prakasam, T.B.S., Srinath, E.G., 1976. Inhibition of nitrification by ammonia and nitrous acid. Journal of the Water Pollution Control Federation 48 (5), 835–852.

Cheng, S.Y., Shieh, L.W., Chen, J.C., 2013. Changes in hemolymph oxyhemocyanin, acid-base balance, and electrolytes in *Marsupenaeus japonicus* under combined ammonia and nitrite stress. Aquatic Toxicology 130–131, 132–138. Available at https://doi.org/10.1016/j.aquatox.2012.12.015.

Clark, E.R., Harman, J.P., Forster, J.R.M., 1985. Production of metabolic and waste products by intensively farmed rainbow trout, *Salmo gaivdnevi* Richardson. Journal of Fish Biology 27 (4), 381–393. Available at https://doi.org/10.1111/j.1095-8649.1985.tb03187.x.

Colt, J., 2006. Water quality requirements for reuse systems. Aquacultural Engineering 34 (3), 143–156. Available at https://doi.org/10.1016/j.aquaeng.2005.08.011.

Davidson, J., Good, C., Welsh, C., Summerfelt, S.T., 2014. Comparing the effects of high vs. low nitrate on the health, performance, and welfare of juvenile rainbow trout Oncorhynchus mykiss within water recirculating aquaculture systems. Aquacultural Engineering 59, 30–40. Available at https://doi.org/10.1016/j.aquaeng.2014.01.003.

Ebeling, J.M., Timmons, M.B., Bisogni, J.J., 2006. Engineering analysis of the stoichiometry of photoautotrophic, autotrophic, and heterotrophic removal of ammonia-nitrogen in aquaculture systems. Aquaculture (Amsterdam, Netherlands) 257 (1-4), 346–358. Available at https://doi.org/10.1016/j.aquaculture.2006.03.019.

Forster, R.P., Goldstein, L., 1969. 5 Formation of excretory products. Fish Physiology 1, 313–350. Available at https://doi.org/10.1016/S1546-5098(08)60086-2.

Fowler, P., Baird, D., Bucklin, R., Yerlan, S., Watson, C., Chapman, F., 1994. Microcontrollers in recirculating aquaculture systems. University of Florida, Florida Cooperative Extension Service. Available at https://www.scribd.com/document/409211260/Microcontrollers-in-Recirculating-Aquaculture-Systems.

Goddek, S., Keesman, K.J., 2018. The necessity of desalination technology for designing and sizing multi-loop aquaponics systems. Desalination 428, 76–85. Available at https://doi.org/10.1016/j.desal.2017.11.024.

Goddek, S., Joyce, A., Kotzen, B., Burnell, G.M., 2019. Aquaponics Food Production Systems: Combined Aquaculture and Hydroponic Production Technologies for the Future. Springer, Cham, pp. 1–620. Available at https://doi.org/10.1007/978-3-030-15943-6.

Good, C., Davidson, J., Welsh, C., Snekvik, K., Summerfelt, S., 2011. The effects of ozonation on performance, health and welfare of rainbow trout Oncorhynchus mykiss in low-exchange water recirculation aquaculture systems. Aquacultural Engineering 44 (3), 97–102. Available at https://doi.org/10.1016/j.aquaeng.2011.04.003.

Haywood, G.P., 1983. Ammonia toxicity in teleost fishes: a review. Canadian Technical Report of Fisheries and Aquatic Sciences 1177, 1–35.

Kamali, S., Ward, V.C.A., Ricardez-Sandoval, L., 2023. Closed-loop operation of a simulated recirculating aquaculture system with an integrated application of nonlinear model predictive control and moving horizon estimation. Computers and Electronics in Agriculture 209, 107820. Available at https://doi.org/10.1016/j.compag.2023.107820.

Lee, P.G., Lea, R.N., Dohmann, E., Prebilsky, W., Turk, P.E., Ying, H., et al., 2000. Denitrification in aquaculture systems: an example of a fuzzy logic control problem. Aquacultural Engineering 23 (1–3), 37–59. Available at https://doi.org/10.1016/S0144-8609(00)00046-7.

Lemarié, G., Dosdat, A., Covès, D., Dutto, G., Gasset, E., Person-Le Ruyet, J., 2004. Effect of chronic ammonia exposure on growth of European seabass (*Dicentrarchus labrax*) juveniles. Aquaculture (Amsterdam, Netherlands) 229 (1–4), 479–491. Available at https://doi.org/10.1016/S0044-8486(03)00392-2.

Li, L., Yakupitiyage, A., 2003. A model for food nutrient dynamics of semi-intensive pond fish culture. Aquacultural Engineering 27 (1), 9–38. Available at https://doi.org/10.1016/S0144-8609(02)00037-7.

Li, T., Lu, J., Wu, J., Zhang, Z., Chen, L., 2022a. Predicting aquaculture water quality using machine learning approaches. Water 14 (18), 2836. Available at https://doi.org/10.3390/w14182836.

Li, W., Wei, Y., An, D., Jiao, Y., Wei, Q., 2022b. LSTM-TCN: dissolved oxygen prediction in aquaculture, based on combined model of long short-term memory network and temporal convolutional network. Environmental Science and Pollution Research 29 (26), 39545–39556. Available at https://doi.org/10.1007/s11356-022-18914-8.

Lin, Y.C., Chen, J.C., 2001. Acute toxicity of ammonia on *Litopenaeus vannamei* boone juveniles at different salinity levels. Journal of Experimental Marine Biology and Ecology 259 (1), 109–119. Available at https://doi.org/10.1016/S0022-0981(01)00227-1.

Liu, Y.T., Wu, C.Y., Chen, T., Yao, Y., 2023. Multi-fidelity surrogate modeling for chemical processes with physics-informed neural networks. Computer Aided Chemical

Engineering 52, 57–63. Available at https://doi.org/10.1016/B978-0-443-15274-0.50010-X.

Markus, M., Hejazi, M.I., Bajcsy, P., Giustolisi, O., Savic, D.A., 2010. Prediction of weekly nitrate-N fluctuations in a small agricultural watershed in Illinois. Journal of Hydroinformatics 12 (3), 251–261. Available at https://doi.org/10.2166/hydro.2010.064.

Panuvatvanich, A., Koottatep, T., Kone, D., 2009. Influence of sand layer depth and percolate impounding regime on nitrogen transformation in vertical-flow constructed wetlands treating faecal sludge. Water Research 43 (10), 2623–2630. Available at https://doi.org/10.1016/j.watres.2009.03.029.

Park, I.S., Lee, J., Hur, J.W., Song, Y.C., Na, H.C., Noh, C.H., 2007. Acute toxicity and sublethal effects of nitrite on selected hematological parameters and tissues in dark-banded rockfish, *Sebastes inermis*. Journal of the World Aquaculture Society 38 (2), 188–199. Available at https://doi.org/10.1111/j.1749-7345.2007.00088.x.

Rychly, J., 1980. Nitrogen balance in trout: II. Nitrogen excretion and retention after feeding diets with varying protein and carbohydrate levels. Aquaculture (Amsterdam, Netherlands) 20 (4), 343–350. Available at https://doi.org/10.1016/0044-8486(80)90095-2.

Tran, H., 2015. Aquaponics-coupled and decoupled systems and the water quality needs of each. 680

Wu, Yan Xiang, Hu, Yong Mei, Liu, Yu Qing, Xue, Min Jie, 2010. The Design of Recirculating Aquaculture Temperature Control System Based on PLC. Applied Mechanics and Materials 40-41, 96–102. Available at 10.4028/www.scientific.net/amm.40-41.96.

Yilmaz, E., 2019. Effects of dietary anthocyanin on innate immune parameters, gene expression responses, and ammonia resistance of Nile tilapia (*Oreochromis niloticus*). Fish & Shellfish Immunology 93, 694–701. Available at https://doi.org/10.1016/j.fsi.2019.08.033.

CHAPTER

Hydroponic planting system

4

Daoliang Li[1,2], Pan Zhang[1,2], Lingwei Jiang[1,3] and Jinqi Yang[3]

[1]*National Innovation Center for Digital Fishery, China Agricultural University, Beijing, P.R. China,*

[2]*College of Information and Electrical Engineering, China Agricultural University, Beijing, P.R. China,*

[3]*School of Chemistry and Chemical Engineering, University of Surrey, Guildford, United Kingdom*

4.1 Vegetable planting system

Hydroponics, initially proposed by W.E. Gericke of the University of California in 1930 (Sheikh, 2006), emerged as a solution for regions facing limited land and water resources. It was envisaged as a method to enhance productivity using fewer fertilizer inputs through modern agricultural practices. The hydroponic system employs three commonly utilized types of hydroponic beds: media-based grow bed, deep water culture (DWC) bed, and nutrient film technique (NFT) gutter-shaped bed (Maucieri et al., 2018), as shown in Fig. 4.1.

The media-based grow bed constitutes a hydroponic trough filled with inert substrate, serving as both root support and microbial substrate. Inert substrates are categorized into Organic substrates (such as peat, coconut fiber, bark, wood fiber) and inorganic substrates (including stone wool, pumice, sand, perlite, vermiculite, expanded clay), as well as synthetic substrates (such as polyurethane, polystyrene). Water is typically supplied in an ebb and flow pattern, ensuring sequential nutrition and aeration (Lewis et al., 1978; Sutton and Lewis, 1982). The DWC system is the cultivation of plants on floating plates or hanging support materials(rafts, panels, boards) in containers filled with 10–30 cm nutrient solution (Van Os et al., 2019). In the DWC system, these plates or materials are generally filled with media, such as rockwool, coco or pumice that support the roots. The NFT consists of narrow channels of perforated square pipes where the roots are partially immersed in a thin layer of streaming water. The NFT system offers several key advantages. First, the recirculation of the nutrient solution and the absence of substrate are among its primary benefits. These features promote efficient nutrient uptake and reduce the risk of soil-borne diseases, respectively. Moreover, the NFT system has significant potential for automation, leading to cost savings in labor for tasks like planting and harvesting, as well as enabling precise control over plant density during the crop cycle. However, it is important to consider the vulnerabilities associated with the NFT system. One of the greatest challenges is

Smart Aquaponic Farming. DOI: https://doi.org/10.1016/B978-0-443-28945-3.00008-6

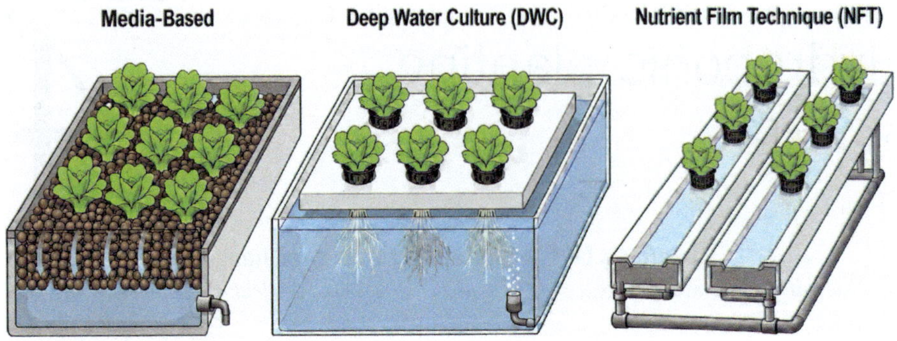

FIGURE 4.1

Scheme of media-based grow bed, deep water culture, and nutrient film technique.

Goddek, S., Joyce, A., Kotzen, B., and Burnell, G.M., 2019 Aquaponics Food Production Systems: Combined Aquaculture and Hydroponic Production Technologies for the Future. *Springer International Publishing, pp. 1–620. https://doi.org/10.1007/978-3-030-15943-6*

that the lack of substrate and low water levels increase the system's susceptibility to pump failures, which may result from issues such as clogging or power supply malfunctions. Consequently, proper maintenance and monitoring are crucial to avoid these complications. In addition, fluctuations in the nutrient solution's temperature can cause plant stress, ultimately leaving the plants more vulnerable to diseases and reducing overall crop health.

4.2 Vegetable growth diagnosis system

4.2.1 Introduction

The aquaponics system combines aquaculture with hydroponic vegetable technology through innovative ecological design, forming a synergistic and complementary circular system (Verma et al., 2023). The core concept of aquaponics is "raising fish without changing water, and planting vegetables without fertilizing", achieving maximum utilization of resources and ecological balance. In the aquaponic system, the role of hydroponic cultivation system is even more indispensable. On the one hand, the feces, urine, and residual bait produced by the fish body lead to the rich nitrogen nutrients required for vegetable growth in the aquaculture wastewater. On the other hand, the roots of vegetables can absorb nutrients from water, purify water quality, and provide a cleaner living environment for fish. This mutually beneficial relationship makes the entire aquaponics system more stable and efficient.

There are quite a variety of vegetables that can be planted in the aquaponics system, mainly including various hydroponic vegetables, which can fully utilize the water resources and nutrients in the system. Common vegetables include water spinach, celery, lettuce, tomatoes, wood ear vegetables, rapeseed, spinach, bok choy, and oilseed rape. These vegetables not only have well-developed root systems and can absorb nutrients from water, but also have a certain purifying effect on water bodies. However, in the actual production process of aquaponics system, due to the asynchronous life cycle of fish and vegetables, dynamic changes in nitrogen nutrient supply in aquaculture wastewater, and differences in nitrogen nutrient requirements of vegetables at different growth stages, it is difficult to ensure optimal crop growth. Therefore, crop growth regulation and growth assessment are particularly important.

At present, the academic community has a rich research foundation in using information technology to monitor and evaluate the growth status of crops or vegetables. Chang et al. (2021) obtained image data from the canopy leaves of melons under different nitrogen concentration treatments, and then used BP, convolutional neural network (CNN), DCNN, and DCNN-LSTM methods to construct leaf nitrogen concentration prediction models. The R^2 of the measured and predicted values of the models were 0.567, 0.376, 0.686, and 0.904, respectively. This study laid the foundation for the rapid, accurate, and intelligent evaluation of nitrogen nutrition in melons. Li et al. (2022) focused on rapeseed in the plant factory scenario and collected image data from three different growth stages of the population. They then constructed a rapeseed target detection model for multiple growth stages using EPSA-YOLOV5s, and embedded the model into a software platform to evaluate the survival rate of rapeseed at different growth stages. Zhao et al. (2022) developed a vegetable disease recognition model DTL-SE-ResNet50 based on artificial intelligence methods. This model is mainly used to identify tomato powdery mildew, tomato leaf mold, and cucumber downy mildew in natural and complex scenes, with an accuracy of 97.24%, which can achieve precise identification of tomato and cucumber diseases. Wang et al. (2024) established a field tea bud dataset and then developed a tea bud target detection model based on improved YOLOv5. Finally, the model was combined with UAV to estimate the number of tea buds before picking and determine the yield. Related research has proven the reliability of using information technology and artificial intelligence technology to solve crop growth monitoring and evaluation in fields or facility crops. However, there is relatively little research on this topic in the aquaponics system. In the following sections, we will focus on discussing the importance, principles, advantages and disadvantages of using information technology and artificial intelligence technology for vegetable growth monitoring and evaluation in the aquaponics system.

4.2.2 Growth monitoring

4.2.2.1 *Nutritional diagnosis*

In the aquaponics system, the nitrogen nutrients required for vegetable growth all come from the aquaculture wastewater. Nitrogen is an essential element for plant growth, and it is the fundamental element that makes up key biomolecules such as plant proteins, nucleic acids, and chlorophyll (Ye et al., 2022). Lack of nitrogen can lead to restricted vegetable growth, yellowing of leaves, decreased photosynthetic efficiency, and ultimately affect overall yield and quality. Excessive nitrogen not only can lead to excessive growth of vegetables, increasing the risk of diseases and pests, but also may affect the quality and taste of vegetables. In the aquaponics, fish excrement can provide nutrients such as nitrogen for vegetables, but it may also produce excessive harmful substances such as ammonia nitrogen. By monitoring the nitrogen nutritional status of vegetables in real-time, accurately, and intelligently, growers can understand the transformation and utilization of nitrogen in the aquaponics system, thereby adjusting the breeding density and plant planting density to maintain the ecological balance of the system. The common diagnostic techniques for vegetable nitrogen nutrition mainly include direct observation, chlorophyll analyzer (soil plant and analyzer development [SPAD]), spectral technology, and visual technology.

4.2.2.1.1 Direct observation

Observing the morphology and leaf color of vegetable plants is the most direct method for diagnosing nitrogen nutrition in vegetables (Li and Li, 2020a). When vegetables experience nitrogen deficiency, they usually exhibit symptoms of slow growth, faded color, and yellowing. Old leaves may first appear yellow, and then gradually develop towards the upper tender leaves, as shown in Fig. 4.2. In severe cases, the entire plant may turn yellow, with small and easily falling leaves. This observation method is simple and intuitive, but requires management personnel to have some planting experience.

4.2.2.1.2 Soil and plant analyzer development

SPAD can quickly and conveniently measure the chlorophyll content in vegetable leaves, indirectly reflecting the nitrogen nutritional status of plants (Yuan et al.,

FIGURE 4.2

Judging the nitrogen nutritional status of vegetables through leaf color. Determine whether crops are nitrogen-deficient by their color.

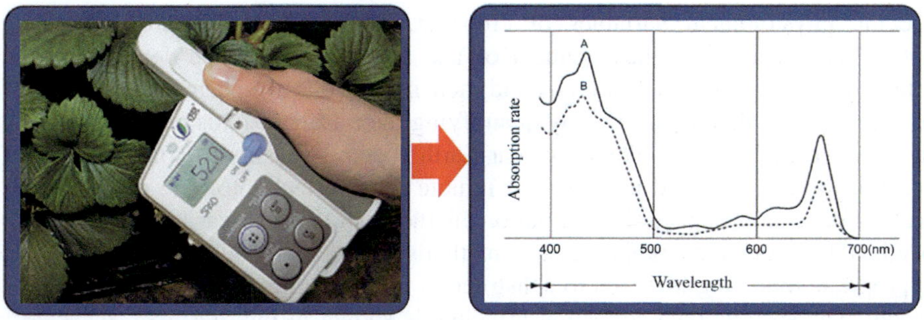

FIGURE 4.3

Soil and plant analyzer development and its measurement principle.

2016). The chlorophyll content is closely related to nitrogen levels, and the SPAD measurement method can monitor the nitrogen uptake dynamics of vegetables in real time, providing a basis for nitrogen fertilization regulation. The SPAD uses photoelectric non-destructive testing method to estimate chlorophyll content by measuring the absorption and reflection characteristics of plant leaves to specific wavelength light. The specific device and principle are shown in Fig. 4.3.

This instrument typically emits two different wavelengths of light sources, one is red light (peak wavelength 650 nm) and the other is near-infrared light (peak wavelength 940 nm). Chlorophyll has a strong ability to absorb red light, while its absorption of near-infrared light is weak. Therefore, by comparing the differences in leaf absorption of these two types of light, the relative content of chlorophyll can be indirectly calculated. Due to the close correlation between chlorophyll content and nitrogen content in leaves, the measurement results of SPAD can reflect the nitrogen nutritional status of vegetables. In general, leaves with higher chlorophyll content indicate sufficient nitrogen supply and vigorous plant growth, Leaves with lower chlorophyll content may have insufficient nitrogen supply, limiting plant growth. Therefore, diagnosing the nitrogen nutritional status of vegetables through SPAD can quickly and conveniently understand the nitrogen demand of vegetables, providing scientific basis for rational fertilization. It should be noted that the measurement results of the SPAD are only relative values and may be affected by factors such as leaf thickness and uneven distribution of chlorophyll. Therefore, other diagnostic methods and field conditions need to be combined for comprehensive judgment in practical applications.

4.2.2.1.3 Spectral analysis

Spectral technology is an important tool for diagnosing the nitrogen nutritional status of vegetables, which can quickly and non-destructively obtain nutritional information of crops. Spectral reflectance is a promising and convenient indicator for continuous sampling and narrow wave selection, which can sensitively reflect specific physiological variables of crops (Li et al., 2020b). In the range of 350–1300 nm, accurate measurement of leaf spectral reflectance depends on the interaction between light and crops and their impact on the spectral characteristics of green leaves, thereby accurately quantifying the nitrogen status of crops. Among them, we usually use handheld or near-Earth spectral sensors to collect crop leaf spectral information, and use UAV or remote sensing technology to collect large-scale crop canopy spectral information in the visible and near-infrared spectral bands. Then, use data preprocessing methods to select feature bands and select appropriate modeling methods to construct an nitrogen nutritional status prediction model. The specific data collection principle is shown in Fig. 4.4.

Nitrogen is the main element that constitutes chlorophyll, therefore, chlorophyll content is closely related to the spectral characteristics of leaves. When the nitrogen supply in vegetables is sufficient, the chlorophyll content in the leaves is high, and the spectral characteristics show enhanced light absorption at certain specific wavelengths; On the contrary, when nitrogen supply is insufficient, chlorophyll content decreases and spectral characteristics also undergo corresponding changes. By collecting and analyzing these spectral data, the nitrogen nutritional status of vegetable leaves can be inferred. This method has the advantages of being non-destructive, fast, and repeatable, and can monitor the nitrogen nutritional status of vegetables in real-time under different growth stages and environmental conditions, providing scientific basis for rational fertilization. In addition, with the continuous

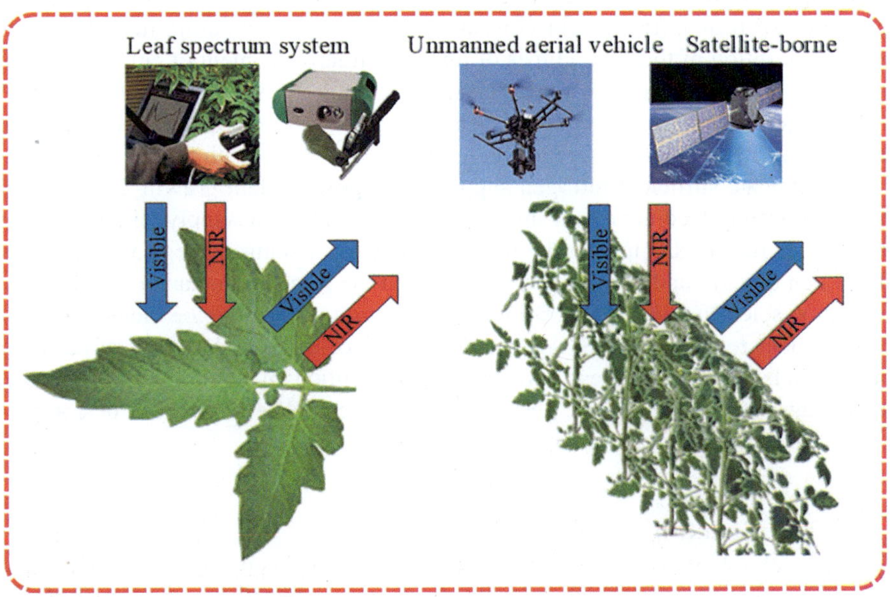

FIGURE 4.4

Spectral technology for vegetable nutrition diagnosis.

development of spectral technology, such as the application of hyperspectral imaging technology, multispectral technology, and so on, we can obtain richer and more accurate spectral information, further improving the accuracy and reliability of vegetable nitrogen nutrition diagnosis. It should be noted that although spectral technology has many advantages, it may still be affected by some factors in practical applications, such as leaf surface structure, water content, and lighting conditions. Therefore, when using spectral technology to diagnose the nitrogen nutritional status of vegetables, it is necessary to comprehensively consider multiple factors and combine other diagnostic methods and techniques for comprehensive analysis and judgment.

4.2.2.1.4 Computer vision

The use of visual technology for diagnosing vegetable nitrogen nutritional status is mainly based on image processing and analysis techniques. This technology collects images of vegetable leaves, extracts feature indicators related to nitrogen nutritional status, and analyzes the relationship between these indicators and nitrogen levels, thereby achieving the diagnosis of vegetable nitrogen nutritional status (Kumar and Bhatia, 2014). The specific acquisition principle is shown in Fig. 4.5.

When collecting images, high-resolution cameras or image capture devices are usually used to ensure that the subtle features of the leaves can be captured. During the

FIGURE 4.5

Application of vision technology in vegetable nutrition diagnosis.

image acquisition process, attention should be paid to factors such as lighting conditions, shooting angle, and distance to ensure that the quality of the collected images meets the analysis requirements. After the image data collection is completed, first, preprocess the collected images, including denoising, enhancing contrast, and other operations, to improve image quality. Second, the leaves are separated from the background through image segmentation technology, laying the foundation for subsequent feature extraction. Finally, feature extraction is performed on the segmented leaf portion. The specific characteristic indicators are introduced as follows:

Color features include RGB values, hue, saturation, and so on of the leaves. When nitrogen supply is sufficient, leaves usually appear bright green, so color characteristics can be an important basis for judging nitrogen nutrition status.

Texture features describe the roughness and particle size of the leaves surface. Insufficient nitrogen may lead to changes in the surface texture of leaves, so texture features are also important diagnostic indicators.

Morphological features include leaf area, perimeter, aspect ratio, and so on. If feature extraction is carried out for whole plants or groups of vegetables, leaf area index and canopy coverage are also important morphological feature indicators. Nitrogen supply has a significant impact on the growth and morphology of leaves, therefore morphological characteristics are also a key indicator for judging nitrogen nutrition status.

With the rise of convolutional neural networks (CNN), image feature extraction can be automatically achieved through CNN. After the relevant feature indicators are extracted, machine learning algorithms (such as support vector machines, neural networks, etc.) are further used to learn and classify the features, thereby establishing a diagnostic model for the nitrogen nutrition status of vegetables. Through this model, vegetable leaf images can be automatically analyzed and diagnosed, and their nitrogen nutritional status can be quickly determined. This method has the advantages of non-contact, fast, and repeatable, and is suitable for practical production environments such as fields or greenhouses. However, it should be noted that visual technology is greatly influenced by factors such as lighting conditions and image quality. Therefore, corresponding measures need to be taken in practical applications to improve the accuracy of diagnosis.

4.2.2.2 Disease diagnosis

The aquaponics system is an interdependent and closely connected ecological cycle system (Fira et al., 2018). In this system, there are complex interactions and energy flows between fish, vegetables, and microorganisms. As a part of the system, the health status of vegetables directly affects the ecological balance of the entire system (Suárez-Cáceres et al., 2021). Once vegetables suffer from diseases, it not only affects their own growth and yield, but may also affect the farmed fish and other organisms through water circulation, thereby disrupting the stability of the entire system. Therefore, vegetable disease diagnosis is a key link in ensuring the ecological balance of the aquaponics system. There are two common causes of vegetable diseases.

1. Biological pathogenicity includes pests and waterborne diseases. Pests can affect the normal growth of vegetables by sucking vegetable juices, spreading diseases, and other means (Fira et al., 2018). For example, pests such as whiteflies, thrips, aphids, cabbage moths, and spider mites can forage and damage vegetables, leading to poor growth or death. Water borne diseases mainly include vegetable diseases caused by Fusarium, Fusarium, and Phytophthora.
2. Non biological pathogenicity mainly includes two aspects: nutritional stress and environmental factors (Fahey et al., 2022). In terms of nutritional stress, calcium is an important component of vegetable cell walls. When the calcium absorbed by vegetables cannot meet their vigorous growth needs, calcium deficiency occurs. It can cause the edges of vegetable leaves to wither and turn yellow, the disease spots to expand, and the leaves gradually dry and turn yellow, ultimately turning black. Iron is an essential trace element in the process of chlorophyll synthesis, which helps maintain the structure of chloroplasts, thereby ensuring the normal greening of vegetable leaves and the progress of photosynthesis. When vegetables are deficient in iron, their new leaves will exhibit yellowing symptoms, with significant loss of green between leaf veins, while the lower leaves remain green. As iron deficiency intensifies, the yellowing phenomenon between leaf veins will shift downward, and young leaves will gradually wither, which may affect root elongation. This iron deficiency symptom is also relatively common in fish vegetable symbiotic systems. Inappropriate environmental factors such as temperature, humidity, and lighting can also lead to vegetable diseases. In addition, if the water quality in the aquaponics system does not meet the standard, it may also cause vegetable diseases.

It should be noted that the diagnosis of vegetable diseases in fish vegetable symbiosis systems usually requires comprehensive consideration of multiple factors, including water quality, nutritional status, environmental conditions, and the growth of the vegetables themselves. Therefore, in practical applications, it is usually necessary to combine multiple diagnostic techniques for comprehensive judgment to improve the accuracy and reliability of diagnosis. At the same time, for the prevention and control of diseases in hydroponic vegetables, attention should also be paid to improving water quality, adjusting nutritional formulas, optimizing environmental conditions, and so on, in order to enhance the disease resistance and yield of vegetables.

4.2.2.2.1 Direct observation

The symptoms of different diseases on hydroponic vegetables vary, so the growth, color changes, and lesion morphology of the roots, stems, and leaves of hydroponic vegetables can be directly observed to distinguish diseases. For example, certain fungal diseases may cause specific shaped and colored spots on leaves, while nutrient deficiency may lead to discoloration or poor growth of leaves. The direct observation method is the simplest and most intuitive diagnostic method, with the advantage of simple and fast operation without the need for special equipment.

FIGURE 4.6

Common cases of lettuce diseases.

However, its disadvantages are also evident, as diagnostic accuracy is greatly influenced by the observer's experience and skills, making it difficult to identify early or asymptomatic diseases as shown in Fig. 4.6.

4.2.2.2.2 Water quality analysis

The growth status of vegetables is closely related to the aquaculture wastewater environment in which they are located. If the water quality of the aquaculture wastewater is not suitable, it may lead to hindered vegetable growth or susceptibility to diseases. Therefore, key parameters such as pH value, dissolved oxygen (DO), and nutrient concentration in aquaculture wastewater can be monitored in real-time through relevant sensors to indirectly reflect the causes of vegetable diseases, as shown in Fig. 4.7. For example, excessive nutrient concentrations may lead to root rot. The water quality analysis method diagnoses diseases by detecting key parameters of aquaculture

FIGURE 4.7

Water quality–related sensors.

wastewater, which can relatively accurately reflect the growth environment of vegetables and help understand the environmental factors that cause diseases. However, changes in water quality parameters may be influenced by multiple factors and may not necessarily directly correspond to specific diseases. Therefore, the accuracy of vegetable disease diagnosis may be limited.

4.2.2.2.3 Microscopic observation method

Part of the diseases in vegetables are caused by microorganisms, which are difficult to distinguish and detect with the naked eye. Therefore, using the high magnification function of the microscope to observe the affected parts of vegetables and identify pathogens such as bacteria, fungi, or parasites, the morphology and structure of the pathogens can be observed, thereby determining the type of disease (as shown in Fig. 4.8). Microscopic examination method diagnoses diseases by observing the morphology and structure of pathogens, which has high accuracy and specificity. However, this method requires professional skills and equipment, and the operation is relatively complex and time-consuming.

4.2.2.2.4 Intelligent recognition method

Artificial intelligence technology has significant advantages in the diagnosis of vegetable diseases in aquaponics systems. First, it can quickly process a large amount of pest and disease image data, achieving automated and efficient

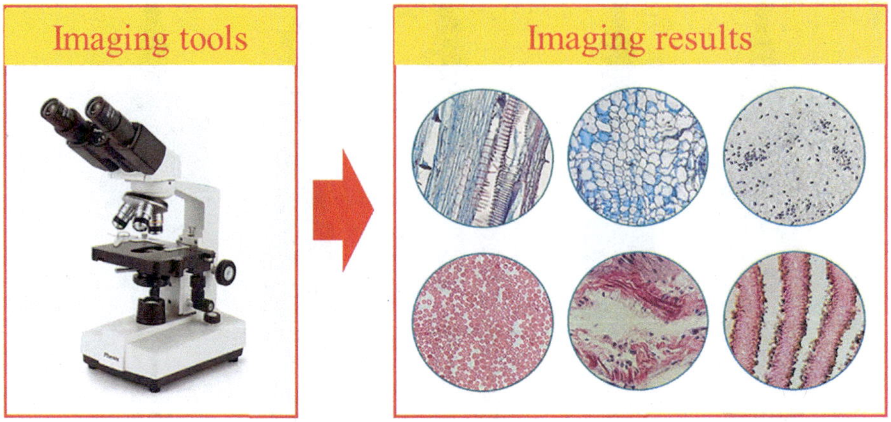

FIGURE 4.8

Microscope and its imaging results.

diagnosis. Second, based on deep learning and machine learning algorithms, artificial intelligence can identify complex and subtle disease symptoms, improving diagnostic accuracy. In addition, artificial intelligence technology can also predict the occurrence trend of diseases based on historical data and real-time environmental information, providing timely and effective decision support for prevention and control work. However, although artificial intelligence technology has obvious advantages, the prerequisite for achieving automatic and accurate vegetable disease recognition is a large amount of annotated data.

4.2.2.2.4.1 Vegetable disease identification based on artificial neural network The artificial neural network (ANN)-based method is mainly aimed at constructing an intelligent recognition model for vegetable disease data with relatively limited data. The disease images of vegetable leaves are collected through devices such as cameras or scanners, and the collected images are preprocessed with denoising, contrast enhancement, color correction, etc. to improve image quality. Then, LBP, GLCM, LBP+GLCM, COLOR, COLOR +LBP, COLOR+GLCM, ALL, SIFT and other methods are used to extract color, texture, shape and other features from the disease image data. The original image is transformed into machine understandable information, so that subsequent classifiers can use this information for disease recognition. Finally, a disease recognition model was constructed using classification methods such as BP neural network, support vector machine, random forest, and k-nearest neighbor. Through this approach, intelligent recognition of vegetable diseases in the fish vegetable symbiotic system was achieved (Anjna Sood and Singh, 2020).

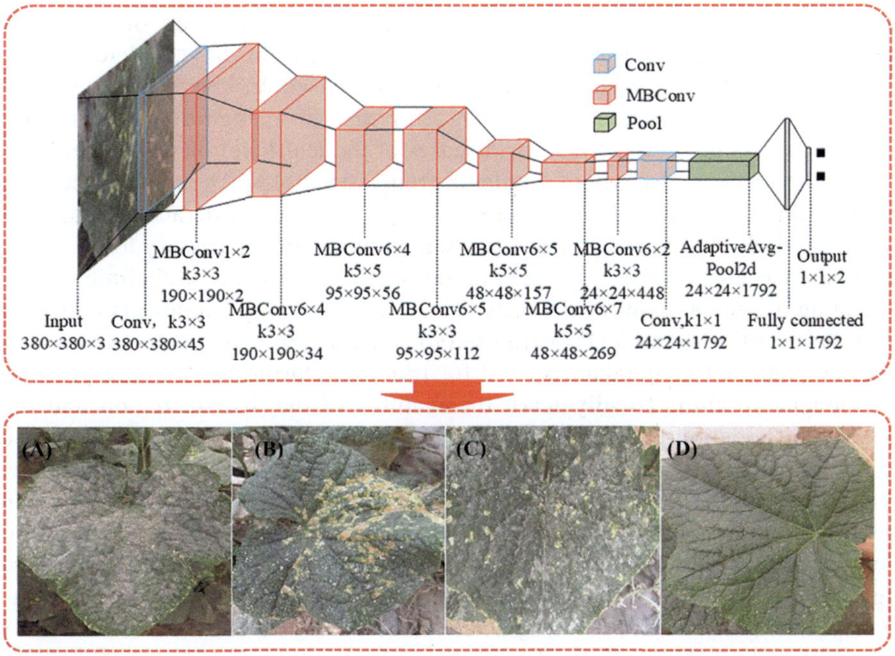

FIGURE 4.9

Identification of vegetable leaf diseases based on EfficientNet-B4.

4.2.2.2.4.2 Vegetable disease identification based on convolutional neural network The vegetable disease recognition based on image classification mainly utilizes the powerful feature extraction and classification capabilities of CNN to learn and analyze the images of vegetable leaves, thereby achieving automatic recognition and classification of vegetable diseases. Unlike vegetable disease recognition methods based on ANN, CNN image classification methods require collecting a large amount of image data containing different types of vegetable diseases for training and optimizing CNN models. CNN models typically consist of multiple convolutional layers, pooling layers, and fully connected layers, such as AlexNet, VGG16, ResNet34, EfficientNet, MobileNet, SqueezeNet, and ShuffleNet. Among them, the convolutional layer is responsible for extracting local features in the image, the pooling layer is used to reduce feature dimensions and computational complexity, and the fully connected layer maps the extracted features to specific disease categories. Taking EfficientNet as an example (Zhang et al., 2020), the specific model construction process is shown in Fig. 4.9.

4.2.2.2.4.3 Vegetable disease recognition based on object detection The occurrence of vegetable diseases in the aquaponics system starts from a local area of the plant, rather than a large area. Therefore, precise identification of early diseases is crucial for preventing large-scale occurrence of vegetable diseases. The vegetable disease recognition based on object detection is different from the vegetable disease recognition based on image classification. The latter is based on the entire image for disease recognition, and mostly constructs classification models for diseases at the leaf level of vegetables. The former can identify diseases of whole or group vegetables. The object detection method can not only achieve disease recognition, but also achieve disease localization. Similar to image classification methods, it is necessary to manually label disease areas using data annotation software (such as Labelimg) when collecting a large amount of datasets. Then, deep learning frameworks such as SSD, Faster RCNN, and YOLO are used to construct an object detection model, and the annotated image data is used for training. During the training process, the model will automatically learn the features of the disease region and the location information of the disease in the image. Based on the trained disease target detection model, when a new vegetable disease image is input, the model will automatically detect and locate the disease spot, while outputting the location information of the disease spot (such as bounding box coordinates) and the type of disease. Li et al. (2022) proposed a vegetable disease detection algorithm based on YOLOv5s, which can be applied to mobile devices such as smartphones and tablets, providing a new theoretical basis and research ideas for disease detection. The specific detection results are shown in Fig. 4.10.

4.2.2.2.4.4 Vegetable disease recognition based on semantic segmentation In addition to identifying and locating diseases, it is also necessary to measure the severity of diseases in the aquaponics system. Image segmentation based methods can not only achieve vegetable disease recognition, but also quantify the degree of disease to a certain extent. Semantic segmentation technology can divide each pixel in an image into different semantic categories, thereby achieving precise segmentation and localization of diseased areas (Lei et al., 2024). Similar to object detection methods, a large number of disease images are collected, and then the disease areas are labeled using labelme software. Based on the annotated image data, a disease recognition model is constructed by substituting it into semantic segmentation networks (such as FCN, UNet, and Swin transformer). By learning the feature representation of the disease area

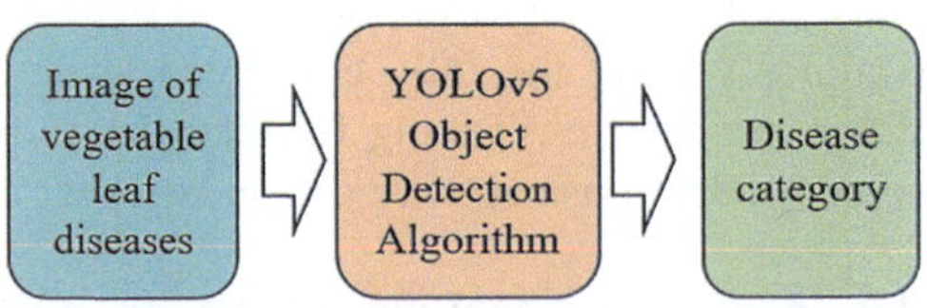

FIGURE 4.10

Vegetable disease recognition based on object detection method.

FIGURE 4.11

Recognition and quantification of vegetable diseases based on semantic segmentation.

through a large amount of training data, the disease information in the image can be automatically extracted and recognized. When inputting new vegetable disease images, the model will automatically segment and locate the diseased areas, and output the ratio of the diseased areas to the overall image to quantify the severity of the disease. Yang et al. (2023) proposed a semantic segmentation model for cucumber leaf disease in complex natural scenes, which can recognize various cucumber leaf diseases, including downy mildew, powdery mildew, target leaf spots, and angular leaf spots The specific results are shown in Fig. 4.11.

4.2.3 Growth assessment

4.2.3.1 Survival rate assessment

In the aquaponics system, the vegetable production process may be affected by factors such as mechanical damage, nutritional stress, or pests and diseases, leading to vegetable death. Survival rate is the basic indicator for evaluating the overall growth status of vegetables in the aquaponics system. By regularly checking and recording the survival rate of vegetables, it is possible to intuitively understand whether the growth environment of vegetables is suitable and whether there are any factors that are not conducive to vegetable growth. In general, when the growth environment of the aquaponics system is suitable and the management level is high, the survival rate of vegetables is usually higher. On the contrary, if the growth environment is harsh or poorly managed, it may lead to a decrease in the survival rate of vegetables. Second, a low survival rate may also indicate issues with water quality, light, temperature, or nutrition, which need to be adjusted in a timely manner. Therefore, it is necessary to evaluate the survival rate of vegetables.

4.2.3.1.1 Manual evaluation

In general, growers at specific stages of vegetable growth, such as after sowing, transplanting, middle growth, and before harvest, directly observe the growth of vegetables, record the number of plants that have withered, died, or grown poorly, and regularly count the number of vegetables. By comparing the number of plants at different time points, the survival rate can be calculated. This method is simple and intuitive, but may be limited by the subjective judgment of the observer. Meanwhile, in large-scale planting scenarios, this method is very time-consuming and may result in significant errors in manual evaluation.

4.2.3.1.2 Intelligent recognition

The evaluation of vegetable survival rate is realized by counting the plants in different growth stages. With the development of artificial intelligence technology, it has become a reality to obtain the image of vegetables in the planting area through the camera, and then call the artificial intelligence model to count the number of plants in different growth stages. This method not only has high real-time performance and accuracy, but also can greatly reduce the time cost of manual evaluation. Among them, vegetable counting can be achieved in low-density planting scenarios such as facilities through methods such as object detection and semantic segmentation. First, by collecting image datasets of vegetables at different growth stages, image annotation is performed and a model training dataset is constructed. Second, target detection or semantic segmentation models are constructed using training datasets, and the reliability of the models is verified. Finally, using Python+PyQt5 to design the software platform, the constructed model is called to achieve the counting of vegetables in the population, and the survival rate of vegetables at different growth stages is evaluated. Zhang et al. (2020) and Li et al. (2022) proposed a rapeseed survival rate evaluation model based on EPSA-YOLOv5s and designed a survival rate monitoring platform for plant factories, as shown in Fig. 4.12.

4.2.3.2 *Yield estimation*

Vegetable yield is one of the key indicators for measuring the efficiency of aquaponics systems. The yield of vegetables directly reflects the production capacity and economic benefits of the system. By comparing the vegetable yields corresponding to different growth stages of aquaculture objects, the main factors affecting vegetable yields can be identified, and the operating mechanism of the system can be optimized to improve the overall yield of the system. Generally, the growth cycle of breeding objects (such as perch) is 8–10 months, and the growth cycle of vegetables (such as lettuce) is generally 1–2 months. In the early stages of fish growth, their food intake is relatively small, their feces and urine are relatively small, and the nitrogen concentration in the aquaculture wastewater is relatively low. As the growth stage of fish increases, the nitrogen concentration in the aquaculture wastewater will gradually increase. The nitrogen nutritional requirements of vegetables vary at different growth stages, and as the growth stage increases, the nitrogen nutritional requirements of vegetables gradually increase. Due to the asynchronous life cycle of fish and vegetables, the dilution effect of nitrogen nutrition supply in the aquaponics system is directly related to the optimal growth of vegetables and the supply of nitrogen nutrition in the aquaculture wastewater. Therefore, real-time and

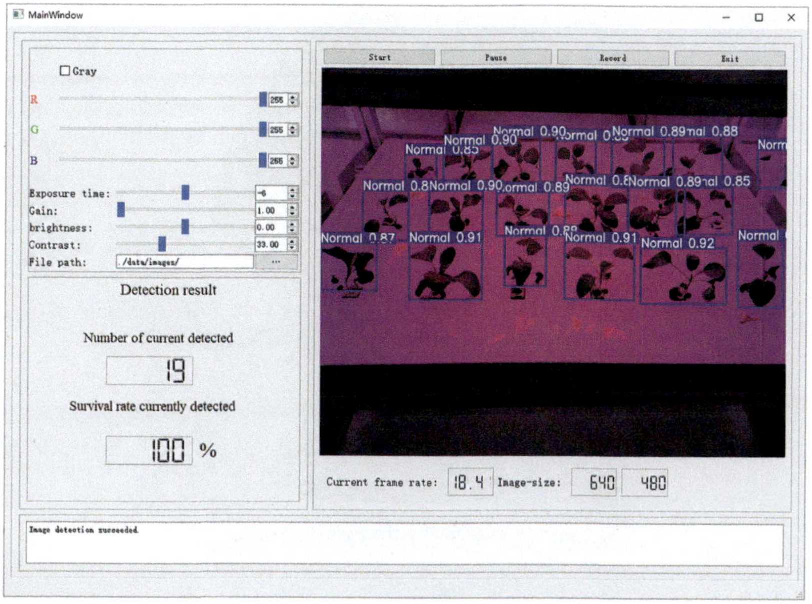

FIGURE 4.12

Vegetable survival rate monitoring platform.

intelligent estimation of vegetable yield is crucial for evaluating the nitrogen utilization efficiency of aquaponics systems.

4.2.3.2.1 Manual measurement

Traditional yield assessment mostly involves rough estimation through direct weighing by growers, which requires vegetables to be picked first and then measured using a balance for yield measurement. Generally, yield evaluation is mainly focused on the harvest stage, and the changes in yield during the growth process are generally not considered. This method can accurately reflect the overall yield of vegetables in the aquaponics system, but it is relatively cumbersome and time-consuming, as shown in Fig. 4.13.

4.2.3.2.2 2D Visual prediction

Visual technology, especially computer vision, can provide powerful data support for yield prediction through real-time monitoring and analysis of vegetable growth status, environmental factors, and plant health status. Among them, two-dimensional vision technology has significant advantages in estimating leaf vegetable yield. First, it is necessary to collect a large amount of vegetable growth image data. These data can be collected through devices such as UAV and cameras, including vegetable images at different growth stages,

FIGURE 4.13

Using a balance for manual weighing of lettuce.

varieties, and environmental conditions. Preprocess these images, such as denoising and enhancing contrast, to improve the accuracy of subsequent image analysis. Second, extract features related to vegetable growth from the preprocessed images, such as leaf color, shape, texture, size, and canopy coverage. These characteristic indicators can reflect the growth and health status of vegetables, and are of great significance for yield prediction. Finally, based on the extracted feature indicators, a yield prediction model is established using machine learning or deep learning regression algorithms. Zhang et al. (2020) used digital images and convolutional neural networks to construct models for predicting the fresh weight, dry weight, and leaf area of lettuce leaves, and the results proved the reliability of CNN for lettuce growth monitoring, the specific network structure is shown in Fig. 4.14.

In practical applications, real-time acquisition of vegetable growth images is used to predict yield using pre-trained prediction models. Based on the predicted results, growers can adjust management measures in a timely manner, such as irrigation and nutrient supply, to optimize the vegetable growth environment and increase yield. It should be noted that vegetable yield prediction based on visual technology is influenced by various factors, such as lighting conditions, image quality, and model complexity. Therefore, it is necessary to comprehensively consider these factors to improve the accuracy and reliability of predictions in practical applications.

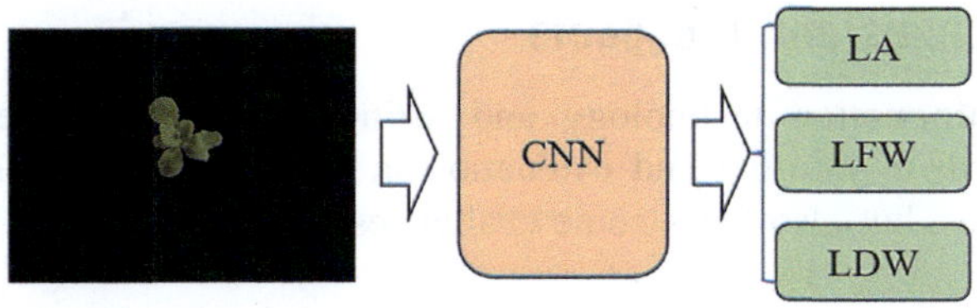

FIGURE 4.14

2D vision technology for predicting lettuce yield.

4.2.3.2.3 3D reconstruction prediction

In addition to planting leafy vegetables, some fruit vegetables will also be planted in the aquaponics system. The growth of fruit vegetables is different from that of leafy vegetables, as they have rich spatial information. When using 3D reconstruction technology for predicting fruit yields, it can provide more accurate and comprehensive plant information, which helps to accurately evaluate the growth status and yield of fruits. First, use high-precision cameras, LiDAR, or other sensors to capture images or point cloud data of fruit from different angles. These data should cover various parts of the plant, including leaves, fruits, and stems. Next, preprocess the data, including noise removal, calibration, and registration, to ensure data quality. Second, based on the obtained multi-view images or point cloud data, a 3D reconstruction algorithm is used to generate a 3D model of fruit and vegetable plants. The reconstructed 3D model can be achieved through techniques such as structured light, stereo vision, or point cloud fusion, and should reflect the shape and structure of the plant as realistically as possible. By extracting features related to fruit growth and yield from 3D models (including plant volume, surface area, number of leaves, fruit size and distribution, etc.) and analyzing them, one can understand the growth status of plants, fruit load conditions, and possible yield trends. Finally, based on the extracted features, combined with historical yield data, environmental factors (such as light, temperature, and humidity), and agricultural operation records, a yield prediction model is established using machine learning or deep learning algorithms. Magistri et al. (2022) trained an encoder using high-resolution unit data, which can predict the 3D shape of fruits online based on a single RGB-D and only requires 4 ms of inference time.

In practical applications, by regularly collecting 3D data of fruit, a pre-trained yield prediction model is used for real-time yield prediction. Based on the prediction results, agricultural producers can timely understand the growth status of plants, formulate targeted management measures, and improve the yield and quality of fruits and vegetables. It should be noted that although 3D reconstruction technology can provide rich plant information, its application is also limited by some factors, such as the cost of data collection equipment, the complexity of data processing, and the impact of environmental factors. Therefore, in practical applications, it is necessary to comprehensively consider these factors in order to fully leverage the advantages of 3D reconstruction technology in predicting fruit and vegetable yields.

4.2.4 Challenges and prospects

The use of information technology and artificial intelligence technology for vegetable growth monitoring and evaluation in aquaponics systems has brought many advantages, but also faces some challenges.

4.2.4.1 Challenges

1. Data integration and standardization: The aquaponics system involves monitoring various parameters and data, such as water quality, temperature, lighting, and plant growth. How to effectively integrate this data and establish a unified data standard is an important challenge.
2. The construction of precise models: To achieve precise monitoring and evaluation of vegetable growth, it is necessary to build accurate artificial intelligence models. However, due to the complexity and dynamics of the aquaponics system, the construction and validation of the model face significant difficulties.
3. The reliability of technology and equipment: The application of information technology and artificial intelligence technology relies on reliable hardware devices and software platforms. In the aquaponics system, these devices and platforms may face harsh environmental conditions, and ensuring their stable operation is a key issue.
4. Data security and privacy protection: With the development of information technology, data security and privacy protection issues are becoming increasingly prominent. In the aquaponics system, how to ensure the security and privacy of data, prevent data leakage and abuse, is an urgent problem that needs to be solved.

4.2.4.2 Prospects

1. Strengthening data analysis and mining: With the development of big data technology, more in-depth analysis and mining of data in aquaponics systems can be carried out in the future to reveal more growth patterns and influencing factors.
2. The development of intelligent decision support systems: By combining artificial intelligence and machine learning technology, intelligent decision support systems can be developed to provide scientific basis for the management and optimization of aquaponics systems.
3. Multidisciplinary integration: The aquaponics system involves multiple disciplines such as agriculture, environment, and computer science. Future research can strengthen the cross integration between these disciplines and jointly promote the development of aquaponics technology.
4. Technology promotion and popularization: By strengthening technology research and demonstration, the threshold and cost of technology application can be lowered, enabling more agricultural practitioners to use information technology and artificial intelligence technology to monitor and evaluate the vegetable growth of aquaponics systems.

4.3 Vegetable growth model and optimization

Over the past several decades, numerous models have been developed to understand and predict plant growth, a process that involves complex and varied biological systems. These models utilize statistical methods to offer reliable predictions for plant growth under typical conditions. For example, dynamic models TOMGRO (Jones et al., 1991), TOMPOUSSE, and TOMSI (Heuvelink, 1999), have significantly enhanced the precision of tomato yield forecasts in greenhouse settings by incorporating various adjustments and integrating with broader climate control systems.

Hydroponic lettuce systems can be configured in various forms, such as DWC or NFT. DWC allows the roots to float in water, while NFT involves a continuous flow of nutrient solution over the roots, which are housed in channels that allow the solution to flow back into the reservoir. Hydroponic lettuce system has been focusing on models aiming to predict the accumulation of dry weight. Innovations like the two-compartment model proposed by Sweeney et al. (Critten, 1991) have provided practical predictions for lettuce cultivation. Modifications to existing models, such as the Nicolet mode (Sweeney et al., 1981), allow predictions of growth rates and nitrate levels in lettuce, improving nutrient management and yield optimization, as shown in Fig. 4.15

Moreover, the growth of plants like lettuce is heavily influenced by environmental factors including temperature, humidity, CO_2 levels, and light intensity. For instance, optimal temperature can enhance the structure of chloroplasts while excessive heat might increase transpiration rates, impacting nutrient uptake negatively. Similarly, CO_2 is crucial for photosynthesis, with deficiencies leading to reduced biomass production.

While these models offer insights into plant responses under various environmental conditions, they are often limited by the availability of comprehensive data on climate and plant responses, which can hinder their application in real-world scenarios. Thus, the development of these models often focuses on a set of specific environmental factors to provide targeted solutions for agricultural production.

4.3.1 Relationship of environment and vegetable growth

Researchers developed a two-compartment model to analyze dry matter accumulation in lettuce (Van Henten, 1994). This model identified economically optimal conditions for air temperature and carbon dioxide levels in the greenhouse environment. It demonstrated that lettuce growth, measured through the dry weight of the crop at harvest, was significantly influenced by the available sunlight (photosynthetically active radiation), ambient temperature, and CO_2 levels. The findings, illustrated in Fig. 4.16, underscored the strong link between outdoor climate conditions, greenhouse management strategies, and economic returns from lettuce sales. These insights reveal how precision in managing environmental factors can directly affect the profitability of crop production.

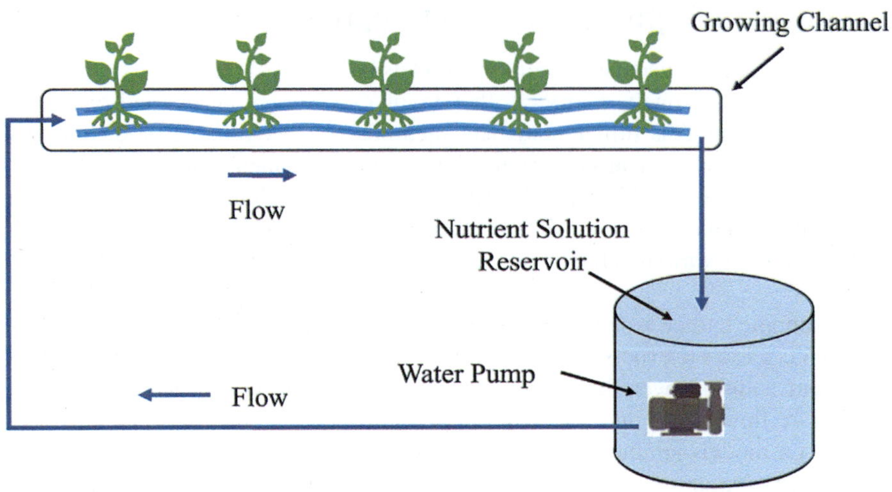

FIGURE 4.15

Hydroponic lettuce system.

Lettuce comprises two main types of carbon storage: structural and non-structural biomass (Goddek et al., 2015). These pools represent the carbon fixed during photosynthesis, which contributes to the plant's physical structure and its energy reserves, respectively. The model outlines how carbon is distributed between these pools, with equations (1)–(10) detailing the specific mechanisms of accumulation. Details on the parameters (shown in Table 4.1) used in these equations can be found in Table 4.2.

4.3.2 Nutrient and plant growth in aquaponics

In aquaponics, maintaining an optimal environment is crucial to prevent pollution and ensure healthy plant growth. Regular monitoring of water quality, including pH, temperature, DO, and nutrient levels like nitrogen, phosphorus, and calcium, is essential. The ideal pH for hydroponics ranges from 5.1 to 6.5, whereas for aquaponics, it should be between 6.5 and 8.0 (Goddek et al., 2015). The recommended temperature for hydroponics is below 70 °F, while for aquaponics, it is typically between 82 and 86 °F.

Research by Yang and Kim highlights the importance of water flow in influencing the quality and distribution of nutrients, which directly affects crop productivity in aquaponic systems (Yang and Kim, 2020). Effective water management is therefore vital. Studies have shown that even small improvements in nutrient availability from solid residues in the water can enhance overall water quality, proving critical for supporting diverse crops such as Tilapia, various leafy greens, and beneficial bacteria necessary for nitrification (Buzby et al., 2016).

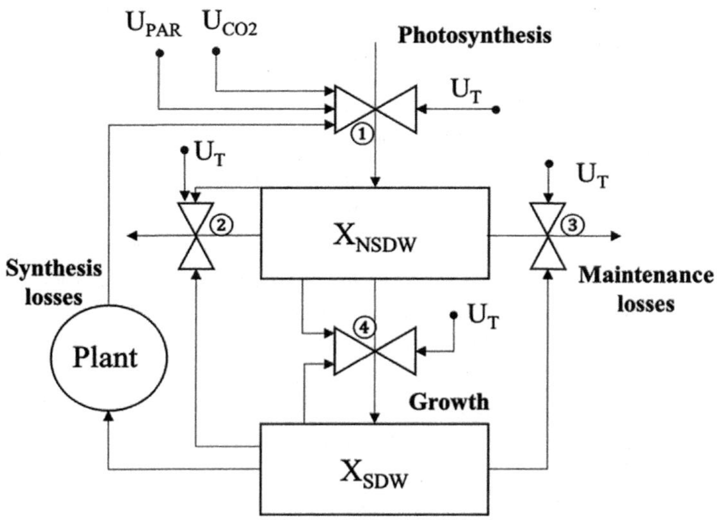

FIGURE 4.16

The dynamic model for lettuce growth.

In addition, advanced techniques like the use of aquaculture wastewater in larger cultivation systems have demonstrated economic and environmental benefits by recycling nutrients. To optimize nutrient levels precisely tailored to plant requirements, researchers like Dhal and colleagues are exploring statistical and machine learning methods (Dhal et al., 2022). These methods aim to refine how nutrients are managed based on dynamic plant needs, fostering the expansion of aquaponics into commercial agriculture. This research has also delved into the effectiveness of both linear and non-linear modeling techniques trained on small datasets to improve decision-making processes regarding nutrient regulation in hydroponic settings, as shown in .

Table 4.3

This rigorous schedule ensured that changes in water chemistry were closely monitored, supporting the maintenance of a stable aquaponic environment.

4.3.3 Challenges and considerations

Despite the successes of well-designed aquaponic systems, significant challenges remain in crop cultivation, particularly concerning nutrient balance. Research suggests that condensed molasses soluble might serve as effective natural fertilizers to overcome these nutrient limitations. In regions like India, Nigeria, and Bangladesh, the aquaculture industry often experiences substantial nitrogen loss, impacting food production security. Aquaponic systems offer a method to analyze nitrogen dynamics through isotopic signatures or stable isotopes, potentially mitigating these losses.

Table 4.1 Parameters in the lettuce growth model.

Parameters	Interpretation
$c_{gr,max}$	Saturation growth rate at 20°C (s^{-1})
$c_{Q10,gr}$	factor for growth
$C_{resp,sht}$	maintenance respiration coefficient for the shoot (s^{-1})
$c_{resp,rt}$	maintenance respiration coefficient for the root (s^{-1})
$c_{Q10,resp}$	Q_{10} factor of the maintenance respiration
c_{τ}	Ratio of the root dry weight to the total crop dry weight
c_K	Extinction coefficient
c_{lar}	Structural leaf area ratio ($m^2 \cdot g^{-1}$)
c_{ω}	Density of CO_2 ($g \cdot m^{-3}$)
c_{Γ}	CO_2 compensation point at 20°C (ppm)
$c_{Q10,\Gamma}$	Effect of temperature on Γ
c_{ε}	Light use efficiency ($g \cdot J^{-1}$)
g_{bnd}	Boundary layer conductance ($m \cdot s^{-1}$)
g_{stm}	Stomatal conductance ($m \cdot s^{-1}$)

In hydroponic systems equipped with artificial lighting, it has been found that nitrogen removal from wastewater is enhanced when exposed to air, especially if the wastewater is also enriched with carbon dioxide. Interestingly, artificial lighting alone does not significantly alter nitrogen levels. In these systems, plants benefit from the nutrients in fish waste, which also serves to purify the water from toxins harmful to fish.

Research indicates that microorganisms that bolster plant growth, such as those associated with sweet peppers, tomatoes, and cucumbers, could explain why aquaponic yields are comparable to those of traditional hydroponics, despite lower nutrient levels. This underscores the need for further study into the effective use of microorganisms across all plant types. In addition, simple electrical conductivity control of nutrient solutions can support sustainable vegetable cultivation using recycled hydroponics.

Soilless cultivation methods are increasingly adopted in protected agriculture to ensure product quality and enhance environmental control, avoiding the variability associated with soil-based nutrients and water. Beyond conserving water, soilless techniques also facilitate urban agriculture by enabling food production on residential rooftops and other urban spaces.

4.4 Nutrient regulation system

4.4.1 Introduction

In the aquaponics system, vegetables require multiple nutrients during their growth process, which play a crucial role in the healthy growth and yield of vegetables. In

Table 4.2 Plant mechanisms of carbon accumulation.

$\frac{dX_{nsdw}}{dt} = c_{\alpha} f_{phot} - r_{gr} X_{sdw} - f_{resp} - \frac{1 - c_{\beta}}{c_{\beta}} r_{gr} X_{sdw}$	(1)
$\frac{dX_{sdw}}{dt} = r_{gr} X_{sdw}$	(2)
$r_{gr} = c_{gr,max} \frac{X_{nsdw}}{c_{\gamma} X_{sdw} + X_{nsdw}} c_{Q10,gr} (U_T - 20)/10$	(3)
$f_{resp} = (c_{resp,sht} (1 - c_{\tau}) X_{sdw} + c_{resp,,t} c_{\tau} X_{sdw}) c_{Q10,resp} (U_T - 25)/10$	(4)
$f_{phot} = (1 - \exp(-c_K c_{lar} (1 - c_{\tau}) X_{sdw})) f_{phot,max}$	(5)
$f_{phot,max} = \frac{\varepsilon U_{PAR} g_{CO_2} c_{\omega} (U_{CO_2} - \Gamma)}{\varepsilon U_{PAR} + g_{CO_2} c_{\omega} (U_{CO_2} - \Gamma)}$	(6)
$\Gamma = c_{\Gamma} c_{Q10,\Gamma} (U_T - 20)/10$	(7)
$\varepsilon = c_{\varepsilon} \frac{U_{CO_2} - \Gamma}{U_{CO_2} + 2\Gamma}$	(8)
$\frac{1}{g_{CO_2}} = \frac{1}{g_{bnd}} + \frac{1}{g_{stm}} + \frac{1}{g_{car}}$	(9)
$g_{car} = c_{car,1} U_T^2 + c_{car,2} U_T + c_{car,3}$	(10)

general, the nutrients required for vegetable growth are mainly divided into large elements, medium elements, and trace elements. Among them, a large number of elements mainly include nitrogen (which has a significant impact on the growth and yield of vegetables and is an important component of protein, nucleic acid, and chlorophyll), phosphorus (which participates in energy transfer and material metabolism of vegetables and plays an important role in root development and fruit formation), and potassium (which plays a key role in maintaining cell osmotic pressure, regulating plant water balance, and enzyme activation). Medium amount elements mainly include calcium (an important component of the cell wall, crucial for maintaining the stability and functionality of cell structure), magnesium (the central atom of chlorophyll, involved in photosynthesis, and has a significant impact on the formation and growth of green leaves in vegetables), and sulfur (although many fertilizer raw materials or finished products contain sulfur, additional sulfur may be needed in certain specific situations to meet the needs of vegetable growth). Trace elements mainly include iron, boron, chlorine, zinc, copper, molybdenum, manganese, and so on. Although these elements have relatively low content in vegetables, they also play an indispensable role in the growth and development of vegetables. They participate in various physiological processes of vegetables, such as enzyme synthesis and activation, electron transfer, and redox reactions.

In the aquaponics system, the aquaculture wastewater (i.e., the water in the fish pond) mainly contains a large amount of elements such as nitrogen, phosphorus, potassium, as well as trace elements such as iron, calcium, magnesium, sulfur, and boron. Among them, nitrogen is the maximum amount of nutrients that fish manure can provide, usually in the form of nitrate to provide nutrients for plants. However,

Table 4.3 Methods used to measure various chemical properties of the aquaponic solution

Property of the aquaponic solution	Process of measurement
pH	Measurement is conducted using a hydrogen ion-selective electrode, which accurately gauges the acidity or alkalinity of the solution.
Conductivity	A conductivity probe is employed to assess the water's ability to conduct electric current, which correlates with the ion concentration in the solution.
Nitrate concentration	Determined through the reduction of nitrite to nitrate using a cadmium column, followed by spectrophotometric measurement to ensure precise quantification.
Chloride concentration	This is Measured through chromatography techniques, which allow for the specific detection and quantification of chloride ions.
Carbonate/Bicarbonate concentrations	These are measured by acid titration using sulphuric acid, a standard method for determining total alkalinity as it reflects the buffering capacity of the water.
Hardness	Calculated based on the concentrations of calcium and magnesium ions, indicating the general mineral content of the water.

due to the varying levels of ammonia nitrogen production in different growth stages of aquaculture objects (fish), the nitrate content converted by nitrifying bacteria is in a dynamic state of change. In addition, the growth cycle of fish and vegetables is not synchronized. The overall nitrogen supply and demand relationship in the fish vegetable symbiotic system is unclear, and it is difficult to ensure optimal crop growth. Therefore, the nutritional regulation of the fish vegetable symbiotic system is of great significance for improving overall economic benefits.

4.4.2 Evidence of nutritional regulation

4.4.2.1 *The main nutritional sources of vegetables*

In the aquaponics system, the aquaculture wastewater is rich in nitrogen element-after nitrification reaction (Wongkiew et al., 2017). Nitrogen is one of the essential elements for vegetable growth, participating in many important biochemical processes within cells. Nitrogen plays a crucial role in the growth process of vegetables. First, nitrogen is a key factor in protein formation in vegetable cells. Proteins are the fundamental substances that make up cells and tissues, and are crucial for the growth and development of vegetables. By applying nitrogen fertilizer reasonably, it can promote the photosynthesis of vegetable leaves, increase the chlorophyll content in the leaves, and thus improve the yield and nutritional value of vegetables. Chlorophyll is a key pigment for vegetable photosynthesis, which can convert light energy into chemical energy and provide energy for vegetable growth. Second, nitrogen also has a significant impact on the root development of vegetables. Moderate nitrogen supply can promote the development of plant roots, improve the absorption capacity of plants for water and nutrients, and further enhance their stress resistance. This is crucial for the survival and growth of vegetables under adverse conditions.

4.4.2.2 *Symptoms of malnutrition in vegetables*

However, in the aquaponics system, there may be some stages (juvenile stage) where the nitrogen supply in the aquaculture wastewater is insufficient to meet the optimal growth of vegetables, which can easily lead to nitrogen deficiency in vegetables. In the previous chapter, we have introduced the technology of vegetable nutrition diagnosis. When nitrogen deficiency occurs, the growth of vegetables will be significantly inhibited. Specifically, the plant is characterized by stunted growth, yellowing of leaves, decreased chlorophyll content, and chlorosis of leaves, especially in old leaves where chlorosis is most pronounced. At the same time, the stems and leaves will become smaller, and the growth of the roots will also be affected. These symptoms can seriously affect the yield and quality of vegetables. If the planting scale is small and the breeding target is adult fish, vegetables may also experience nitrogen excess. In general, excess nitrogen can lead to an increase in the number of branches and leaves in hydroponic vegetables, resulting in elongation. This will make the plant less robust and affect its normal growth. At the same time, due to excessive nitrogen, the number of vegetables flowering will decrease, the fruit setting rate will decrease, the fruit may appear deformed, and the quality will also become inferior. On the contrary, when the nitrogen fertilizer content in the tail water is too high, it can lead to a concentration of aquaculture tail water greater than that of root hair cell fluid. In this case, the water in the root hair cell fluid will seep into the aquaculture wastewater, causing the root hair cells not only to be unable to absorb water, but also to lose water. Vegetable wilting occurs due to the inability to absorb sufficient water from the aquaculture wastewater, which is known as seedling burning. Therefore, it is crucial to control the supply of nitrogen fertilizer reasonably during vegetable cultivation. It is necessary to ensure that vegetables obtain sufficient nitrogen to meet their growth needs, while avoiding the negative effects caused by excessive application. Through scientific nitrogen supply management, the high-yield, high-quality, and sustainable development of vegetables in the fish vegetable symbiotic system can be achieved.

In general, leafy vegetables are the main species in the aquaponics system. The main edible part of leafy vegetables is their leaves, which grow rapidly and require a large amount of nitrogen to promote chlorophyll synthesis and protein formation. Therefore, leafy vegetables require relatively more nitrogen fertilizer. During the later stage of nutritional growth, especially during the differentiation of flower buds and fruit enlargement, although fruits and vegetables also have a high demand for nitrogen, their nutrient uptake is not the highest throughout the entire growth cycle. Moreover, compared to leafy vegetables, fruit vegetables do not have the most significant demand for nitrogen. Therefore, in the actual nutritional regulation process of fish vegetable symbiosis system, it is necessary to regulate according to the nutritional requirements of different vegetables.

4.4.3 Nutritional regulation based on optimal nitrogen migration and transformation

In aquaponic systems, biofilters play a crucial role in maintaining water quality and providing essential nutrients for plant growth by converting toxic ammonia (NH_3) into harmless nitrate (NO_3^-). Biofilters can be categorized into fixed bed biofilters (filled with plastic rings or ceramic rings), fluidized bed biofilters (utilizing suspended sand or plastic particles), moving bed biofilm reactors (equipped with floating plastic sheets or rings), and trickling filters (where water flows over packed media towers). Each type offers a high surface area for microbial attachment, ensuring efficient oxygen transfer and nitrogen conversion. The system comprises biological media, water flow and aeration systems, solid waste separation devices, and control and monitoring equipment to optimize operation and stabilize water quality.

Fixed bed biofilters are filled with high-surface-area fixed media, such as plastic rings or ceramic rings, providing ample surface for microbial attachment and growth. These biofilters are simple in structure and easy to maintain but are prone to clogging by solid waste, necessitating regular cleaning.

Fluidized bed biofilters use lightweight materials such as sand or plastic particles that remain suspended in the water flow, offering higher surface area and better oxygen transfer efficiency. These are suitable for large-scale aquaponic systems, but the system complexity, high initial investment, and need for specialized maintenance knowledge can be challenging.

Moving bed biofilm reactors contain floating biological carriers (usually specially designed plastic pieces or rings) that move continuously in the water flow, combining the benefits of fixed and fluidized beds. They feature strong self-cleaning capabilities, resistance to clogging, and low maintenance costs, but require stable water flow and an appropriate power system to keep the carriers in motion.

Trickling filters operate by spraying water over a tower packed with biological media (such as plastic, stone, or gravel), where gravity causes the water to flow through the media. This design allows for natural air circulation, providing ample oxygen. Trickling filters are efficient in oxygen transfer and nitrogen conversion, suitable for medium to large systems, but require substantial installation space and high construction costs.

In biofilters, the conversion of ammonia nitrogen to nitrate nitrogen is influenced by several factors, including pH, DO, hydraulic loading rate (HLR), and carbon-to-nitrogen ratio (C/N ratio). pH is a crucial factor in controlling fish metabolism, microbial activity, and plant nitrogen uptake, with nitrifying bacteria showing optimal activity at a pH range of 6.4–9.0. Dissolved oxygen is essential for maintaining the activity of nitrifying bacteria and other aerobic microorganisms; DO levels below 1.7 mg/L inhibit the activity of AOB and NOB, with a recommended DO concentration of 5–6 mg/L to prevent stress on fish and plants. Hydraulic loading rate, defined as the liquid flow rate per unit area of the grow bed, optimizes nutrient contact with plant roots and microorganisms. Both excessively low and high HLR can impact oxygen transfer and microbial activity in the system. A high C/N ratio promotes the growth of heterotrophic bacteria, reducing the number and efficiency of nitrifying bacteria, and leading to nitrogen loss.

4.4.4 Nutrition regulation based on production process

As mentioned earlier, we introduced the nutritional diagnosis of vegetables and the nitrogen migration and transformation in aquaculture wastewater. The nutritional diagnosis of vegetables mainly provides a basis for nitrogen nutrition regulation in aquaponics systems. By exploring the nitrogen migration and transformation laws in aquaculture wastewater, better regulation and optimal transformation of nitrate nitrogen can be achieved. In addition, the optimization of the aquaponics system at the process level can also achieve optimal nutritional regulation. The Chongqing Academy of Agricultural Sciences in China has proposed a "one main and two auxiliary" planting and breeding cycle system for fish and vegetable symbiosis. This system can ensure that the nitrate nitrogen content of the entire fish and vegetable symbiosis system is less than 500 mg/L, ammonia nitrogen is less than 0.5 mg/L, and nitrite nitrogen concentration is less than 0.2 mg/L. Ensuring the quality of vegetable cultivation and achieving high-density fish farming. The system mainly includes three fish tail water circulation routes, namely the main circulation route, the first auxiliary line, and the second auxiliary line. The operation process of the entire system is shown in Fig. 4.17.

4.4.4.1 The main circulation route

The main circulation route includes fish ponds, microfiltration machine, fluidized beds, ozone disinfection devices, and ultraviolet disinfection devices. The upper drainage pipeline of the fish pond is connected to the microfilter. After being filtered by the microfilter, the drainage enters the fluidized bed, where ammonia nitrogen and nitrite nitrogen are converted into nitrate nitrogen, and then returns to the fish pond through ozone disinfection and ultraviolet disinfection devices.

4.4.4.2 The first auxiliary line

The first auxiliary line includes fish ponds, microfiltration machine, fluidized beds, vegetable cultivation systems, filters, ozone disinfection devices, and ultraviolet disinfection devices. A branch pipe is separated from the outlet of the fluidized bed and connected to the vegetable cultivation system. The outlet of the vegetable cultivation system is connected to the filter, and the outlet of the filter is connected to the ozone disinfection device. The water from the fluidized bed enters the vegetable cultivation system, where it absorbs nitrate nitrogen from the aquaculture wastewater and reduces its content. After being filtered by the filter, the water from the vegetable cultivation system is returned to the fish pond through the ozone disinfection device and ultraviolet disinfection device.

4.4.4.3 *The second auxiliary line*

The second auxiliary line includes a fish pond, a vertical flow sedimentation tank, a fish manure collection tank, a fish manure thickener, an anaerobic fermentation tank, an aerobic fermentation tank, and a symbiotic system of bacteria and algae. The bottom drainage pipeline of the fish pond is connected to the vertical flow sedimentation tank, and the clear liquid outlet pipeline of the vertical flow sedimentation tank is connected to the microfilter. The fish manure from the vertical flow sedimentation tank and the fish manure filtered out by the microfilter enter the fish manure collection tank. The filtrate outlet pipeline of the microfilter divides into a branch pipe and is connected to the fish manure collection tank. The fish manure from the fish manure collection tank is concentrated by the fish manure concentrator and enters the anaerobic fermentation tank for anaerobic fermentation. Then, it enters the aerobic fermentation tank for aerobic fermentation. After aerobic fermentation, the tail water enters the bacteria and algae symbiosis system again. The water from the bacteria and algae symbiosis system then passes through the ozone disinfection device and ultraviolet disinfection device before returning to the fish pond.

Under normal circumstances, the main circulation route, the first auxiliary line, and the second auxiliary line operate simultaneously to meet the needs of fish farming, as well as the cultivation of fungi and vegetables. At the same time, fish manure can be treated. When the nitrate nitrogen content of the water from the fluidized bed is between 500 mg/L and 1000 mg/L, most of the water from the fluidized bed is allowed to enter the vegetable cultivation system to further remove ammonia nitrogen through vegetable cultivation, ensuring that the nitrate nitrogen content in the water returning to the fish pond is below 500 mg/L to adapt to the growth of fish. When the nitrate nitrogen content is greater than 1000 mg/L, the water from the microfilter is allowed to enter the third auxiliary line for anaerobic reaction to increase nitrogen removal.

4.4.5 Challenges and prospects

In the fish vegetable symbiotic system, nutrient regulation at the vegetable planting end is a complex and critical process that involves multiple factors and challenges.

4.4.5.1 *Challenges*

First, the nutritional regulation of vegetable cultivation in the fish vegetable symbiosis system needs to ensure that vegetables obtain sufficient nutrients while

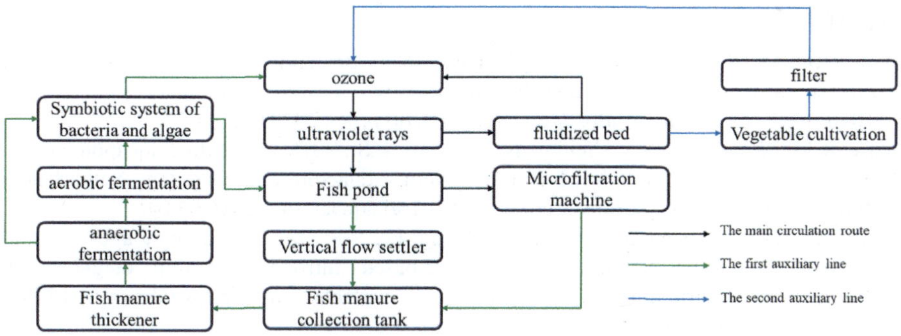

FIGURE 4.17

A planting and breeding cycle system with "one main and two auxiliary."

avoiding potential environmental pollution. However, due to the dynamic changes in water nutrients in the aquaponics system, precise regulation of the nutrient levels required for vegetables has become particularly difficult. In addition, different types of vegetables have different nutritional requirements, which makes nutritional regulation more complex.

Second, the migration and transformation of nitrogen in water is a key link in nutrient regulation. In the aquaponics system, organic substances such as fish excrement are decomposed by microorganisms and converted into inorganic nitrogen, which is absorbed and utilized by vegetables. However, this process is influenced by various factors, such as water temperature, light, microbial activity, etc. The changes in these factors may lead to fluctuations in nitrogen conversion efficiency, thereby affecting the growth and quality of vegetables.

4.4.5.2 Prospects

In response to the above challenges, future research should focus on the following aspects: (1) in-depth study of the mechanism of nitrogen migration and transformation in aquaponics systems, revealing its influencing factors and regulatory pathways, and providing theoretical support for optimizing nutritional regulation. (2) Develop an intelligent nutrition regulation system that monitors the content and changes of nutrients in water in real-time, automatically adjusts fertilization strategies, and achieves precise fertilization. (3) Strengthen the selection and adaptability research of vegetable varieties, screen suitable vegetable varieties for aquaponics system cultivation, and improve the efficiency of vegetable utilization of nutrients; (4) Promote interdisciplinary cooperation and exchange, draw on research achievements and methods from other disciplines, and jointly promote the development and innovation of fish vegetable symbiosis technology.

References

Anjna Sood, M., Singh, P.K., 2020. Hybrid system for detection and classification of plant disease using qualitative texture features analysis. Procedia Computer Science 167, 1056–1065. Available at https://doi.org/10.1016/j.procs.2020.03.404.

Buzby, K.M., Waterland, N.L., Semmens, K.J., Lin, L.S., 2016. Evaluating aquaponic crops in a freshwater flow-through fish culture system. Aquaculture (Amsterdam, Netherlands) 460, 15–24. Available at https://doi.org/10.1016/j.aquaculture.2016.03.046.

Chang, L., Li, D., Hameed, M.K., Yin, Y., Huang, D., Niu, Q., 2021. Using a hybrid neural network model DCNN–lSTM for image-based nitrogen nutrition diagnosis in muskmelon. Horticulturae 7 (11). Available at https://doi.org/10.3390/horticulturae7110489.

Critten, D.L., 1991. Optimization of CO_2 concentration in greenhouses: A modelling analysis for the lettuce crop. Journal of Agricultural Engineering Research 48 (C), 261–271. Available at https://doi.org/10.1016/0021-8634(91)80020-f.

Dhal, S.B., Bagavathiannan, M., Braga-Neto, U., Kalafatis, S., 2022. Nutrient optimization for plant growth in aquaponic irrigation using machine learning for small training datasets. Artificial Intelligence in Agriculture 6, 68–76. Available at https://doi.org/10.1016/j.aiia.2022.05.001.

Fahey, C., Koyama, A., Antunes, P.M., 2022. Vulnerability of non-native invasive plants to novel pathogen attack: do plant traits matter? Biological Invasions 24 (11), 3349–3379. Available at https://doi.org/10.1007/s10530-022-02853-z.

Fira, D., Dimkić, I., Berić, T., Lozo, J., Stanković, S., 2018. Biological control of plant pathogens by Bacillus species. Journal of Biotechnology 285, 44–55. Available at https://doi.org/10.1016/j.jbiotec.2018.07.044.

Goddek, S., Delaide, B., Mankasingh, U., Ragnarsdottir, K.V., Jijakli, H., Thorarinsdottir, R., 2015. Challenges of sustainable and commercial aquaponics. Sustainability (Switzerland) 7 (4), 4199–4224. Available at https://doi.org/10.3390/su7044199.

Heuvelink, E., 1999. Evaluation of a dynamic simulation model for tomato crop growth and development. Annals of Botany 83 (4), 413–422. Available at https://doi.org/10.1006/anbo.1998.0832.

Jones, J.W., Dayan, E., Allen, L.H., Van Keulen, H., Challa, H., 1991. Dynamic tomato growth and yield model (TOMGRO). Transactions of the American Society of Agricultural Engineers 34 (2), 663–672.

Kumar, G., Bhatia, P.K., 2014. A detailed review of feature extraction in image processing systems. 2014 Fourth International Conference on Advanced Computing & Communication Technologies, Rohtak, India. IEEE, pp. 5–12. Available at 10.1109/ACCT.2014.74.

Lei, L., Yang, Q., Yang, L., Shen, T., Wang, R., Fu, C., 2024. Deep learning implementation of image segmentation in agricultural applications: a comprehensive review. Artificial Intelligence Review 57 (6). Available at https://doi.org/10.1007/s10462-024-10775-6.

Lewis, W.M., Yopp, J.H., Schramm, H.L., Brandenburg, A.M., 1978. Use of hydroponics to maintain quality of recirculated water in a fish culture system. Transactions of the American Fisheries Society 107 (1), 92–99. Available at https://doi.org/10.1577/1548-8659(1978)107[92:UOHTMQ]2.0.CO;2.

Li, D., Li, C., Yao, Y., Li, M., Liu, L., 2020a. Modern imaging techniques in plant nutrition analysis: a review. Computers and Electronics in Agriculture 174. Available at https://doi.org/10.1016/j.compag.2020.105459 105459.

Li, D., Zhang, P., Chen, T., Qin, W., 2020b. Recent development and challenges in spectroscopy and machine vision technologies for crop nitrogen diagnosis: a review. Remote Sensing 12 (16), 2578.. Available at https://doi.org/10.3390/rs12162578.

Li, J., Qiao, Y., Liu, S., Zhang, J., Yang, Z., Wang, M., 2022. An improved YOLOv5-based vegetable disease detection method. Computers and Electronics in Agriculture 202, 107345. Available at https://doi.org/10.1016/j.compag.2022.107345.

Magistri, F., Marks, E., Nagulavancha, S., Vizzo, I., Läbe, T., Behley, J., et al., 2022. Contrastive 3D shape completion and reconstruction for agricultural robots using RGB-D frames. IEEE Robotics and Automation Letters 7 (4), 10120–10127. Available at https://doi.org/10.1109/LRA.2022.3193239.

Maucieri, C., Nicoletto, C., Junge, R., Schmautz, Z., Sambo, P., Borin, M., 2018. Hydroponic systems and water management in aquaponics: A review. Italian Journal of Agronomy 13 (1), 1–11. Available at https://doi.org/10.4081/ija.2017.1012.

Sheikh, B.A., 2006. Hydroponics: Key to Sustain Agriculture in Water Stressed and Urban Environment. Pakistan Journal of Agriculture, Agricultural Engineering and Veterinary Sciences 22, 53–57.

Suárez-Cáceres, G.P., Pérez-Urrestarazu, L., Avilés, M., Borrero, C., Lobillo Eguíbar, J.R., Fernández-Cabanás, V.M., 2021. Susceptibility to water-borne plant diseases of hydroponic vs. aquaponics systems. Aquaculture (Amsterdam, Netherlands) 544. Available at https://doi.org/10.1016/j.aquaculture.2021.737093.

Sutton, R.J., Lewis, W.M., 1982. Further observations on a fish production system that incorporates hydroponically grown plants. The Progressive Fish-Culturist 44, 55–59. Available at https://doi.org/10.1577/1548-8659(1982)44[55:FOOAFP]2.0.CO;2.

Sweeney, D., Hand, D.W., Slack, G., Thornley, J.H.M., 1981. Modelling the growth of winter lettuce. Mathematics and plant physiology. Academic Press, pp. 217–229.

Van Henten, E.J., 1994. Validation of a dynamic lettuce growth model for greenhouse climate control. Agricultural Systems 45 (1), 55–72. Available at https://doi.org/10.1016/s0308-521x(94)90280-1.

Van Os, E.A., Gieling, T.H., Lieth, J.H., 2019. Technical equipment in soilless production systems. Soilless Culture: Theory and Practice Theory and Practice. Elsevier, Netherlands, pp. 587–635.

Verma, A.K., Chandrakant, M.H., John, V.C., Peter, R.M., John, I.E., 2023. Aquaponics as an integrated agri-aquaculture system (IAAS): emerging trends and future prospects. Technological Forecasting and Social Change 194. Available at https://doi.org/10.1016/j.techfore.2023.122709.

Wang, S.M., Yu, C.P., Ma, J.H., Ouyang, J.X., Zhao, Z.M., Xuan, Y.M., et al., 2024. Tea yield estimation using UAV images and deep learning. Industrial Crops and Products 212. Available at https://doi.org/10.1016/j.indcrop.2024.118358.

Wongkiew, S., Hu, Z., Chandran, K., Lee, J.W., Khanal, S.K., 2017. Nitrogen transformations in aquaponic systems: a review. Aquacultural Engineering 76, 9–19. Available at https://doi.org/10.1016/j.aquaeng.2017.01.004.

Yang, R., Guo, Y., Hu, Z., Gao, R., Yang, H., 2023. Semantic segmentation of cucumber leaf disease spots based on ECA-SegFormer. Agriculture 13 (8), 1513. Available at https://doi.org/10.3390/agriculture13081513.

Yang, T., Kim, H.J., 2020. Effects of hydraulic loading rate on spatial and temporal water quality characteristics and crop growth and yield in aquaponic systems. Horticulturae 6 (1). Available at https://doi.org/10.3390/horticulturae6010009.

Ye, J.Y., Tian, W.H., Jin, C.W., 2022. Nitrogen in plants: from nutrition to the modulation of abiotic stress adaptation. Stress Biology 2 (1). Available at https://doi.org/10.1007/s44154-021-00030-1.

Yuan, Z., Cao, Q., Zhang, K., Ata-Ul-Karim, S.T., Tan, Y., Zhu, Y., et al., 2016. Optimal leaf positions for SPAD meter measurement in rice. Frontiers in Plant Science 7 (2016). Available at https://doi.org/10.3389/fpls.2016.00719.

Zhang, L., Xu, Z., Xu, D., Ma, J., Chen, Y., Fu, Z., 2020. Growth monitoring of greenhouse lettuce based on a convolutional neural network. Horticulture Research 7 (1). Available at https://doi.org/10.1038/s41438-020-00345-6.

Zhang, P., Yang, L., Li, D., 2020. EfficientNet-B4-ranger: a novel method for greenhouse cucumber disease recognition under natural complex environment. Computers and Electronics in Agriculture 176, 105652. Available at https://doi.org/10.1016/j.compag.2020.105652.

Zhao, X., Li, K., Li, Y., Ma, J., Zhang, L., 2022. Identification method of vegetable diseases based on transfer learning and attention mechanism. Computers and Electronics in Agriculture 193, 106703. Available at https://doi.org/10.1016/j.compag.2022.106703.

CHAPTER

Air Environment Control System 5

Daoliang Li[1,2], Xiaoyang Wu[3] and Jinqi Yang[3]
[1]*National Innovation Center for Digital Fishery, China Agricultural University, Beijing, P.R. China,*
[2]*College of Information and Electrical Engineering, China Agricultural University, Beijing, P.R. China,*
[3]*School of Chemistry and Chemical Engineering, University of Surrey, Guildford, United Kingdom*

5.1 Outdoor environment monitoring system

The vegetables and fish are grown in a relatively stable indoor environment, the microclimate, which is significantly affected by the natural environment, such as outdoor temperature, sunlight, and wind speed. The weather station can be calibrated to measure the outdoor air temperature and humidity, wind speed, wind direction, rainfall, and outdoor soil. An outdoor environment monitoring system (Fig. 5.1) is a comprehensive setup designed to continuously observe and analyse various environmental parameters in outdoor settings (Zhang et al., 2022). These systems are vital for aquaponics production. Here's a breakdown of the key components and functionalities:

Sensors: These are devices responsible for measuring different environmental parameters. Depending on the application, sensors may include:

Weather sensors: air temperature, humidity, air pressure, wind speed, wind direction, rainfall, and solar radiation.

Air quality sensors: Monitor levels of pollutants such as particulate matter (PM2.5, PM10), ozone (O_3), nitrogen dioxide (NO_2), sulfur dioxide (SO_2), carbon monoxide (CO), and volatile organic compounds (VOCs).

Data acquisition and transmission: Sensor data is collected and transmitted to a central processing unit (CPU) or cloud-based server for analysis and storage. This process may involve wired connections (e.g., Ethernet) or wireless protocols such as Wi-Fi, LoRa WAN, or cellular networks.

Central processing unit: The CPU processes incoming data, executes algorithms for analysis, and generates actionable insights. It may also manage device configurations, data storage, and communication protocols.

User interface: A user-friendly interface allows users to visualise real-time and historical data collected by the system. This can take the form of web-based dashboards, mobile applications, or desktop software. Users can monitor environmental conditions, set alerts for abnormal readings, and access reports for analysis.

Smart Aquaponic Farming. DOI: https://doi.org/10.1016/B978-0-443-28945-3.00005-0

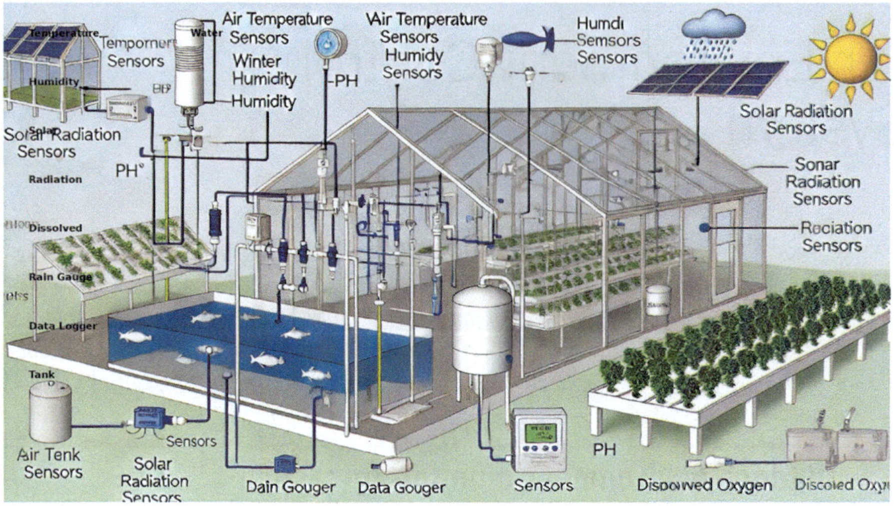

FIGURE 5.1

Details of the air environment control system of the smart aquaponics system.

No permission required.

Alerting and notification system: The system can be configured to send alerts or notifications to relevant stakeholders when predefined thresholds are exceeded or when anomalies are detected. This enables timely responses to environmental emergencies or regulatory compliance issues.

Integration with geographic information systems (GIS): Integration with GIS platforms allows for spatial analysis and visualisation of environmental data. This enables users to overlay sensor data with geographic features such as maps, terrain, land use, and infrastructure, facilitating better decision-making and resource allocation.

Environmental modelling and forecasting: Advanced systems may incorporate environmental modelling and forecasting capabilities. By analysing historical data and current conditions, these systems can predict future trends and potential environmental impacts, helping stakeholders proactively plan and respond to changing conditions.

Energy optimisation and sustainability: Outdoor monitoring systems can help identify opportunities for energy optimisation and sustainability initiatives. For example, by adjusting lighting and heating systems in response to weather conditions.

Figure 5.1 shows the details of the air environment control system of the smart aquaponics system.

5.2 Indoor spatial parameter monitoring system

5.2.1 Introduction

The traditional aquaculture faces the resource consumption and environmental pollution problems (Bergland et al., 2020). Many researchers have explored the mechanisms of in-situ restoration and ex-situ restoration, and continue to improve the breeding model focused on the water quality (Pedersen and Wik, 2020). Now the cleaner aquaponics in greenhouse has become more and more popular in the present agriculture (Yep and Zheng, 2019). The stability of the environment directly affects production. The stability of the environment directly affects the production; it is necessary to explore a stable monitoring system to observe the breeding environment.

In recent years, highly intelligent monitoring technologies have been developed in agricultural systems due to the use of new information and communication technologies. ZigBee has implied the integration of communication capabilities to sensors (Sun and Zhang, 2018), which can track the environmental parameters in real time. Yang (2020) elaborated on the progress of agricultural Internet of Things technology in aquaponics. Deng designed an intelligent control aquaponics system including automatic data acquisition, recording, analysis and control (wei et al., 2020; Deng et al., 2020). The intelligent monitoring system in the aquaponics is also highly popular with foreign researchers. Witzel proposed an integrated hydroponic production system control and regulation, management execution system method (Witzel et al., 2019). Tolentino introduced a monitoring and automatic correction system for a hydroponic system building in a temperature-controlled greenhouse, which was based on the Internet of Things (IoT) using Android devices (Tolentino et al., 2019). The system mainly includes the collection of real-time data monitoring from the light intensity sensor and the air temperature and humidity sensor. This focused on the monitoring of the pH and temperature of the system's circulating water and plant canopy. To solve the identified problems in traditional hydroponics, Menon proposed a technique that uses wireless sensors and a communication network with GUI applications to solve the identified problems in traditional hydroponics (Menon, 2020). An indoor spatial parameter monitoring system is a valuable tool for aquaponics to ensure indoor environments are conducive to productivity while optimising energy usage and operational efficiency.

5.2.2 Environment parameters

The parameters related to the air conditions in contact with the plants are analysed in this section. To obtain and maintain balanced and safe optimal crops in the system and to ensure the stable and healthy growth of fish and vegetables, it is necessary to monitor and control some environmental parameters. Namely, the air temperature, where the ranges change between the plant species; the light intensity, which depends on the growth stage of the plant; the air humidity, the air content of CO_2 (carbon dioxide), and the media moisture, in case of adopting a media base as the hydroponic component.

5.2.2.1 ***Air temperature***

The air temperature influences the health of plants. The suitable temperature for most of the vegetables commonly grown in aquaponics systems is 18°C–30°C. At higher temperatures, leafy greens bolt and begin to flower and seed. Further, the air temperature is responsible for a correct transpiration of the crops. The factors that have been taken into consideration when selecting air temperature sensors are: (1) temperature range; (2) contact or contactless; (3) sensing element; and (4) calibration method. For measuring the air temperature in aquaponics systems, a thermistor has been the recurrent option, as the air temperature and humidity are measured together.

5.2.2.2 ***Relative humidity***

The relative humidity is an expression for moisture in the air. Most of the grown crops in aquaponics systems need humid air for thriving, thus the relative humidity needs to be well managed. The humidity in the air can be measured in different ways: (1) mass of water in unit volume of air; (2) unit mass of air; or (3) partial pressure of water vapour in the air. Air humidity can also be expressed as a proportion of the air–water saturation or relative humidity (RH).

Warm air has a higher moisture-holding capacity than cooler air. To maintain control of the level of humidity, good control over ventilation and heating systems is necessary. The ventilation system allows the exchange of the moisture inside the greenhouse with drier air from outside the facilities. The heating devices are necessary to warm up the outdoor air to the optimum growing temperature and increase the capacity of the air to hold moisture. To increase moisture in indoor air, it is common to use moisturisers. The optimum level of RH varies depending on the type of crop and the growth stage of the plants. The most common considerations are 50%–80%, but those depend on the indoor temperature. An excess of RH in the air would interfere with the plants' transpiration and prevent roots and stems from supplying an adequate quantity of water to the leaves. As mentioned earlier, air RH is usually provided by the air temperature sensor.

5.2.2.3 ***Carbon dioxide***

Carbon dioxide is an essential component of photosynthesis, a vital chemical reaction for plants to sustain. In mass production indoor systems, plants may use all the CO_2 available in the air. As such, artificial addition of CO_2 and control of its levels are necessary. The optimum range level for most of the indoor crops is within 340–1300 ppm. However, a smaller range may be needed for some crops. For example, 800–1000 ppm of CO_2 in the air is necessary to grow tomatoes, cucumbers, peppers, or even lettuce. Depending on the type of crop, the lighting conditions, air temperature and RH, different CO_2 air concentrations are needed.

5.2.2.4 *Light intensity*

Sunlight is critical for plants, but is unavailable or limited in indoor facilities. Artificial lighting is placed in aquaponics systems as a substitute for natural light to provide light to the plants. Light is usually measured in terms of its intensity (lux). However, plants use a limited part of the light spectrum called photosynthetically active radiation (PAR). It designates the spectral range of solar radiation that photosynthetic organisms can process.

5.2.3 System components

An indoor spatial parameter monitoring system is a sophisticated network of sensors, software, and hardware designed to monitor and manage various environmental conditions within indoor spaces. These systems are essential for ensuring optimal conditions for aquaponics. Here's a breakdown of its components and functionalities:

Sensors: These are the backbone of the system, capturing data on parameters such as temperature, humidity, air quality (CO_2 levels, VOCs, particulate matter), light intensity, noise levels, occupancy, and even movement patterns. Sensors can be deployed strategically throughout the indoor environment to provide comprehensive coverage.

Data acquisition and transmission: Sensors collect real-time data and transmit it to a CPU or a cloud-based server for analysis and storage. Communication methods can include wired connections or wireless protocols like Wi-Fi, Zigbee, or Bluetooth.

Central processing unit: The CPU processes incoming data, executes algorithms for analysis, and generates actionable insights. It may also manage device configurations, data storage, and communication protocols. In some cases, edge computing techniques are employed to perform initial data processing locally before transmitting it to a central server.

User interface: A user-friendly interface, often accessible through a web-based dashboard or mobile application, allows users to visualise real-time and historical data collected by the system. Users can monitor environmental conditions, set alerts for abnormal readings, and access reports for performance analysis.

Alerting and notification system: The system can be configured to send alerts or notifications to designated personnel when predefined thresholds are exceeded or when anomalies are detected. This ensures timely responses to issues such as temperature fluctuations, poor air quality, or overcrowding.

Integration with building management systems: In commercial or institutional buildings, integration with BMS allows for centralised control and automation of HVAC (Heating, Ventilation, and Air Conditioning), lighting, security, and other building systems based on real-time data collected by the monitoring system.

Energy optimisation and sustainability: By continuously monitoring environmental parameters and analysing energy consumption patterns, these systems help identify opportunities for energy optimisation and sustainability initiatives. For example, HVAC systems can be adjusted based on occupancy levels to minimise energy waste while maintaining comfort.

Compliance and reporting: Indoor spatial monitoring systems often include features for compliance tracking and reporting, particularly in regulated environments such as healthcare facilities or laboratories. Automated reporting functionalities streamline compliance documentation and auditing processes.

5.2.4 Scheme design

The real-time monitoring and management are the main parts of the aquaponics management system. The IoT technology was employed in the management system performance, including the real-time environmental monitoring, remote control, over-limit alarm, video monitoring, and other functions for the planting, which greatly reduced labour intensity, saved management costs and improved production efficiency. In general, the demonstration sites consist of: vegetable growth environment, fish growth water and outside climate, the monitoring system must keep the reliability, practicability and accuracy for the three parts. The aquaponics environmental monitoring system was divided into three parts as follows: monitoring of the fish and vegetables, the microclimate environment in the greenhouse and the outdoor environment. Considering the diversity of environmental parameters, it is necessary to development suitable system for realising the changes in the growth environment of fish and vegetables. This section focuses on the monitoring of indoor spatial environment parameters.

The monitoring system was divided into a perception layer, a transmission layer and an application layer (Fig. 5.2). The details are as follows:

Perception layer: It mainly includes water quality monitoring nodes, weather monitoring nodes, hydroponic environment nodes and on-site sensor networks. The real-time data collected on-site is transmitted upward through the use of RS485 and LOR combined communication methods.

Transmission layer: The data from the first layer, which will be sent to the cloud platform through the 4G network, Javer, was employed to establish the surface.

Application layer: The function of the application layer is to preprocess data and forward it to the user. It could not only monitor the data on the platform but also complete data storage and dynamic display. The functions of the monitoring system include: (1) Real-time monitoring of the temperature and humidity of hydroponic plant leaves. (2) Real-time monitoring of the indoor temperature, humidity and carbon dioxide concentration.

5.2.5 Hardware platform and Internet of Things architecture

IoT communication technologies are employed to connect sensors and the server. The investment cost of the system is completely within the scope of farmers'

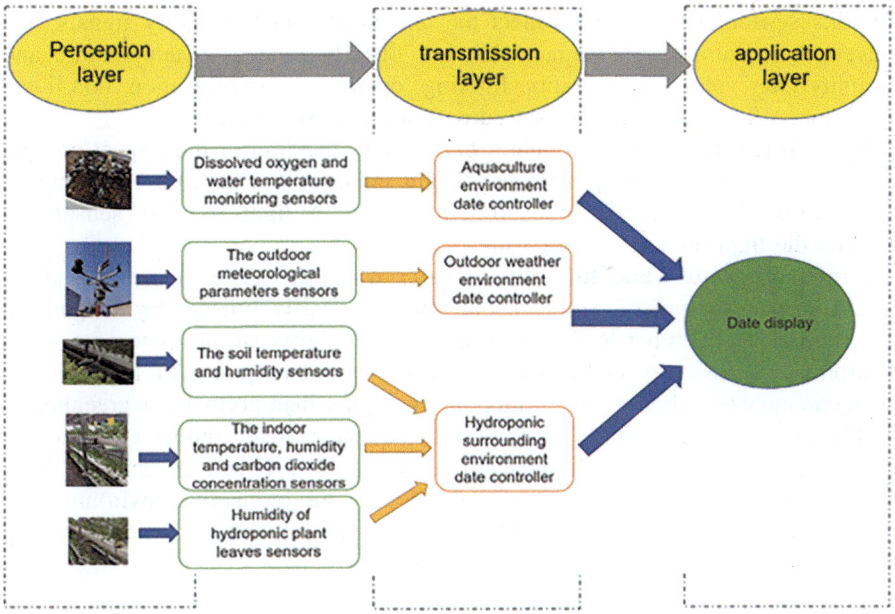

FIGURE 5.2

The monitoring system scheme design diagram.

Van Beveren, P.J.M., Bontsema, J., Van Straten, G., Van Henten, E.J., 2015. Minimal heating and cooling in a modern rose greenhouse. Applied Energy 137, 97–109. https://doi.org/10.1016/j.apenergy.2014.09.083.

acceptance. According to the requirements of hardware equipment for product durability, simple operation principle, reasonable price, and convenient purchase, the system uses as much general equipment as possible.

For the stable hydroponics environment in a recirculating aquaponics system included: the photosynthesis of vegetables, the oxygenation of the bubbles generated by the aerator, vegetable respiration and transpiration, light intensity and indoor ventilation. A variety of sensors can be employed to reflect environmental conditions, such as a carbon dioxide concentration sensor, a light sensor, a soil temperature and humidity sensor, air temperature and humidity sensor. The carbon dioxide produced by fish through respiration also affects the photosynthesis of vegetables. The system was equipped with multiple sets of carbon dioxide sensors at different spatial points to monitor the distribution of carbon dioxide in the hydroponic area in real time.

Temperature and humidity sensors: The growth of plants has strict requirements for temperature and humidity. Under improper temperature and humidity, plants will stop growing or even die. The system can employ the temperature and humidity sensor, which could collect the temperature and humidity signals, and convert them into current signals that are linearly related to temperature and humidity after voltage stabilisation filtering, operational amplification, non-linear correction, V/I conversion, constant current and reverse protection.

Light intensity sensor: To gain a better understanding of the impact of light intensity conditions on the greenhouse. Light intensity can be affected by indoor temperature, photosynthesis and soil temperature. A light intensity sensor can measure the light intensity in real time.

Leaf temperature and humidity sensor: When the leaves of the vegetables have a certain amount of water vapour, they are easily infected by fungal and bacterial diseases. Proper leaf temperature and humidity can promote crop water absorption, improve crop cell metabolism, and promote crop growth. Especially in the aquaponics greenhouse, the air humidity is higher than normal in a greenhouse.

The plant growth cycle is an important indicator of the output efficiency of aquaponics in a greenhouse. Exploring the correlation between the environmental changes and growth period could not only monitor the production environment in real time, but also calculate the growth cycle and environmental changes according to the video record.

5.3 Indoor air environment measurement and control system

5.3.1 Greenhouse microclimate mathematical model

To improve system understanding of the greenhouse, mechanism models with an emphasis on physical processes have been studied during the past few decades (López-Cruz et al., 2018).

In research on thermal models for controlling the microclimate within greenhouses, Sethi and colleagues categorized these models into three types: independent, heating-dependent, and cooling-dependent (Sethi et al., 2013). Cunha reviewed a variety of approaches for simulating greenhouse microclimates, encompassing physically-based models, black-box linear parametric models, and black-box non-linear parametric models (Cunha, 2003). These efforts aim to optimize and manage the microclimates in greenhouse. López-Cruz showed the value of mechanistic and black-box models. They also made the case that dynamic models need sensitivity analysis and uncertainty analysis to be more reliable. Artificial neural networks (ANN) could fit the complicated system that could not be solved by a physical-based model (Iddio et al., 2020; Escamilla-García et al., 2020).

Fig. 5.3 illustrates an input-output transmission system of the greenhouse aquaponics. In this system, we can manipulate window, illustration, and heating to adjust microclimate, and keep air temperature, inside humidity and CO_2 and so on. In the control process, we should consider various disturbances, such as wind speed, water temperature solar radiation, and outside environment factors, which will have influence on the states of greenhouse microclimate.

Because of the characteristics of strong coupling, nonlinearity, and strong disturbance, the greenhouse modeling is complex. Therefore, several optimization techniques and intelligent learning methods have been promoted to improve greenhouse modeling performance. Such as Multi-layer Perceptron, ANN, Support Vector Machine, and Multivariate Adaptive Regression Splines, and so on. The machine learning and heuristic optimization algorithms improved the nonlinear fitting ability of models to complex systems (Mehdizadeh, 2018; Mohammadi et al., 2021). These various optimization and reinforcement techniques have been thoroughly investigated to build a model that will operate more consistently and accurately for complicated agricultural greenhouse systems.

5.3.2 Dynamic optimal and control method in greenhouse

A predictive control approach, constrained by a model, was formulated utilizing a discrete-time nonlinear representation of the greenhouse environment. This kind of model incorporated various states such as the temperatures of the air, crops, floor, and greenhouse cover. It accounted for disturbances like both direct and diffuse

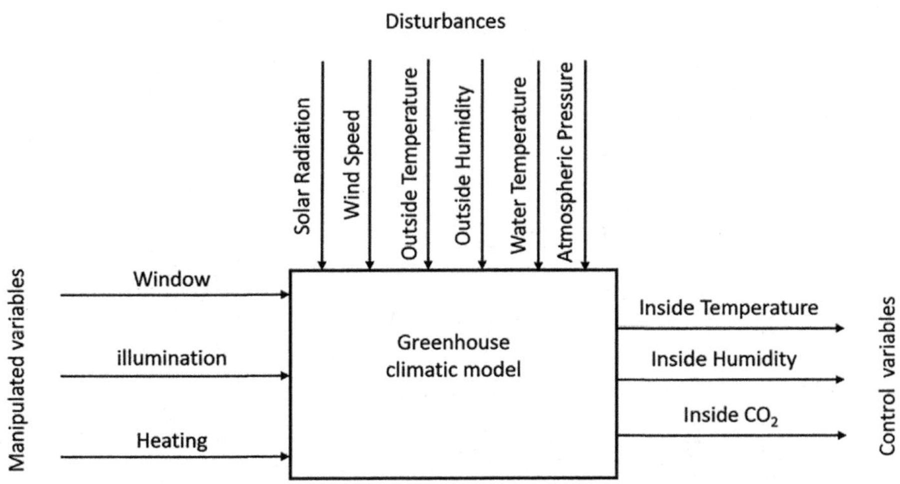

FIGURE 5.3

Greenhouse aquaponics input–output transmission system.

From Blasco, X., Martínez, M., Herrero, J.M., Ramos, C., Sanchis, J., 2007. Model-based predictive control of greenhouse climate for reducing energy and water consumption. Computers and Electronics in Agriculture, 55(1), 49–70. https://doi.org/10.1016/j.compag.2006.12.001.

Table 5.1 Key characteristics of the dynamic greenhouse climate models.

Model/ Author	Control inputs	Type	Calibration	Evaluation	SA	Greenhouse
Van Henten	CO_2, Heating, Window	ODEs	Y	Y	Local	Glasshouse
Tap	CO_2, Heating, Window	ODEs	Y	Y	Local	Glasshouse
Trigui	Heating, humidification, Window	ODEs	Y	Y	Preliminary	Glasshouse
Pohlheim	CO_2, Heating, ventilation, vapor	ODEs	Y	Y	N	Glasshouse
Solaer-1	Valve position of heating and cooling	ODEs	Y	Y	Local	Glasshouse
Van Ooteghem	Heating pump, heat exchanger, ventilation with heat recovery, window,	ODEs	Y	Y	N	Glasshouse
Van Beveren	CO_2, heating, cooling, ventilation	ODEs	Y	Y	N	Glasshouse
Coelho	Heating, Window	Arx	Y	Y	N	Polyethylene GH
Blasco	Heating, Fogging, Window	ODEs	Y	Y	N	Polyethylene GH

ODEs, ordinary differential equations; *DE*, difference equations; *Arx*, autoregressive with external inputs; *NE*, Not specified; *SA*, Sensitivity analysis; *GH*, greenhouse; *Y*, include; *N*, Not include.

Iddio, E., Wang, L., Thomas, Y., McMorrow, G., Denzer, A., 2020. Energy efficient operation and modeling for greenhouses: a literature review. Renewable and Sustainable Energy Reviews, 117. https://doi.org/10.1016/j.rser.2019.109480.

solar radiation, external temperature, and the soil temperature at 10 cm depth. The regultion of heat via heating or cooling systems represented the primary controllable factor. The model should also take into account the essential physical processes occurring within the greenhouse, such as energy transform and energy balance.

In a greenhouse, controlling the internal environment effectively requires advanced models that consider various factors such as air temperature, carbon dioxide levels, and absolute humidity (Piñón et al., 2005). One common approach is to use dynamic models that provide a framework for predictive control, which adjusts the internal conditions based on real-time data and predictive algorithms.

For instance, autoregressive models (Table 5.1) with external inputs can predict future air temperatures by analysing the differences in current internal and external temperatures, the amount of solar radiation, and the heat from the greenhouse's heating system. These models adjust their predictions based on inputs from heating and ventilation controls, ensuring that the greenhouse environment remains optimal for plant growth.

Furthermore, the optimization of water and energy usage within the greenhouse can be enhanced through models that utilize genetic algorithms (Blasco et al., 2007). These algorithms help in fine-tuning the greenhouse's responses to external weather conditions by adjusting factors such as window openings, fogging systems, and heating settings based on predicted environmental changes.

The physical and thermal dynamics within the greenhouse are also crucial. Models need to account for heat transfer processes including conduction, convection, and the heat generated by crop transpiration. Moreover, managing humidity through calculated ventilation and fogging can prevent excess moisture and ensure a balanced environment for plants.

Overall, the integration of smart control systems and predictive models can significantly improve the efficiency of greenhouse operations, reducing energy consumption while maximizing agricultural output. These models continually adjust to external disturbances such as changes in sunlight, wind speed, and outdoor temperatures, ensuring that the internal conditions are always conducive to plant health and productivity (Table 5.1).

5.3.3 Model predictive control

MPC, developed in the 1970s from heuristic control methods, is a versatile and widely applicable control system, commonly used in various complex industrial processes and now being adapted for greenhouse management (Mayne et al., 2000). This control method functions by forecasting and optimizing control actions based on current system data, continually adjusting to achieve the best immediate outcomes.

MPC, also referred to as rolling horizon or receding horizon control, operates by continuously updating its predictions at each interval, using real-time data to recalibrate and optimize the control process. This method contrasts with traditional control strategies by focusing on a dynamic approach that adjusts based on ongoing data analysis rather than static preset parameters.

The system relies on three core components (Fig. 5.4):

1. **Predictor**: Utilizes historical data and current conditions to forecast future states of the system under different control settings.
2. **Cost function**: Assesses the potential cost of each action, guiding the system towards economically efficient decisions.
3. **Optimization method**: Continuously seeks the most effective action by solving complex mathematical problems in real-time, ensuring optimal performance.

Through these mechanisms, MPC offers a significant advantage in real-time optimization, making it an essential tool for managing environments where conditions change rapidly, such as in greenhouses.

The controlled system can be approximated as a difference equation:

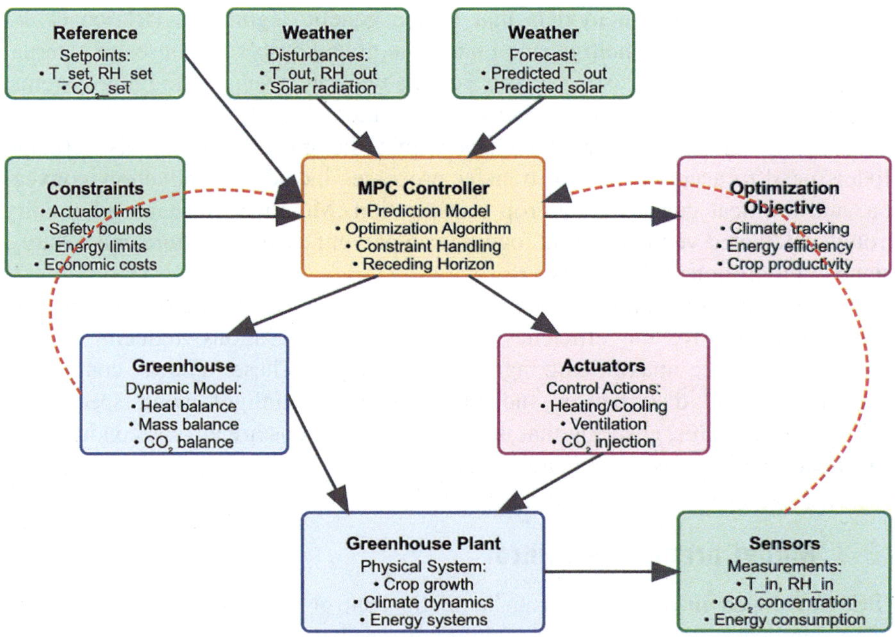

FIGURE 5.4

Greenhouse model predictive control basic elements.

Van Beveren, P.J.M., Bontsema, J., Van Straten, G., Van Henten, E.J., 2015. Minimal heating and cooling in a modern rose greenhouse. Applied Energy, 137, 97–109. https://doi.org/10.1016/j.apenergy.2014.09.083.

$$x^{+}\& = f(x, u) \tag{5.1}$$

$$y\& = h(x) \tag{5.2}$$

where $x \in \mathbb{R}^n$ is the state quantity of the system, u is the control quantity, x^+ is the state quantity of the system at the next moment; $f(x, u)$ is the system state transition function, which has a balance point at $f(0, 0) = 0.u$ is the finite control sequence $\{u(k), u(k+1).\cdots,u(k+N-1)\}$, the corresponding system state sequence x is $\{x(k+1), x(k+2), \cdots, x(k+N)\}$, where k is the current time. The control and state sequence of the system must satisfy the following conditions:

$$u(k) \in U \tag{5.3}$$

$$x(k) \in X \tag{5.4}$$

where $\mathbb{U}$ is a convexly compact subset of $\mathbb{R}^m$, x^+ is a convexly closed subset of $\mathbb{R}^n$, all sets include the origin. The control objective of the model predictive controller is to drive the system state to the origin or equilibrium point x_s, while outputting $y_s = h(x_s) = r$, where the target output r is a constant.

Define the objective function $V(\cdot)$ as the objective function of the system, for the moment k, the initial state of the system is x, the finite-time optimal control problem of model predictive control optimization $\mathbb{P}_N(x)$ is defined as:

$$V_N^0(x) = \min_u \{V_N(x, u) | u \in U_N(x)\} \tag{5.5}$$

where $\mathcal{U}_N(x)$ is the set of admissible control sequences satisfying state constraints, control quantity constraints and terminal constraints, which can be expressed as:

$$U_N(x) = u \in U^N \mid x^u(i) \in X, u(i) \in U, i \in I_(k: k+N-1), x^u(k+N) \in X_f \tag{5.6}$$

where x^u is the state of the system under the control quantity $\boldsymbol{u}$. The objective function V_N: $\mathbb{R}^n \times \mathbb{U}^N \rightarrow \mathbb{R}$ can be defined as:

$$V_N(x, u) = \sum_{i=k}^{k+N-1} \sum l(x(i),u(i)) + V_f(x(k+N)) \tag{5.7}$$

In MPC, the stage objective function $l(x, u) \geq 0$. If the origin is the equilibrium point of the system, the phase objective function can be written as:

$$l(x, u)\& = x'Qx + u'Ru \tag{5.8}$$

$$V\_f(x)\& = x'P_f x \tag{5.9}$$

where Q, R are the tuning parameters in the controller, and P_f is the weight matrix of the terminal penalty. Compared with R, a large value of Q will quickly drive the system to the origin, but will result in a larger change in the control quantity. When the value of R is relatively large, the control quantity will be reduced, but the time for the system to reach the steady state point will be longer. Q and P_f are positive semi-definite, R is positive definite.

The solution to the optimization problem is the optimal control sequence $u^0(x) = \{u^0(k;x), u^0(k+1;x), \cdots, u^0(k+N-1;x)\}$, the corresponding optimal state sequence is $x^0(x)$. In the control sequence $u^0(x)$, only the first quantity $u^0(k;x)$ is applied to the practical system. Therefore, the control rate of MPC is:

$$\kappa_N(x) = u^0(k; x) \tag{5.10}$$

The number of controllable variables is fixed. The optimization time domain N of the optimal control problem solved online should be infinite. In practice, N is usually limited. The stability and recursive feasibility of the algorithm are usually achieved in two different ways: the first way adds a terminal penalty function $V_f(\cdot)$ and a terminal constraint set $\mathbb{X}_f$ to the objective function. The other way by choosing a long enough time domain N. Both them have rich research results and theories (Rawlings and Muske, 1993).

5.3.4 Optimal control of greenhouse climate using minimal energy

In regions with temperate climates, such as the Netherlands, greenhouse crop production significantly relies on fossil fuels. Recognizing the environmental impact of this reliance, the Dutch government alongside local agricultural producers has set specific goals to reduce energy use and lower carbon emissions. To achieve these objectives, a new strategic approach has been developed. This approach focuses on enhancing energy efficiency in greenhouses by optimizing the use of resources and reducing unnecessary energy expenditures, all while ensuring that the conditions (Fig. 5.5) within the greenhouses remain ideal for crop growth (Van Beveren et al., 2015).

Greenhouse climate control involves managing temperature, humidity, and CO_2 concentration to ensure optimal crop growth. Traditional methods rely heavily on heuristic rules and the grower's experience, leading to inefficiencies in energy use. There is a need for a more systematic approach to reduce energy consumption without compromising crop yields.

Researchers from Wageningen University developed a dynamic model to control greenhouse climate using minimal energy. The model incorporates:

- Temperature and humidity control
- CO_2 concentration management
- Optimization of heating, cooling, natural ventilation, and CO_2 injection

The study used data from a 40,709 m² Venlo-type greenhouse in Bleiswijk, Netherlands, over the year 2012. The researchers employed optimal control techniques, which do not require detailed crop production models but focus on minimizing resource input based on grower-defined climate bounds.

These results were achieved by dynamically adjusting the greenhouse climate settings within predefined bounds. The model's effectiveness was validated through comparisons with actual data from the greenhouse:

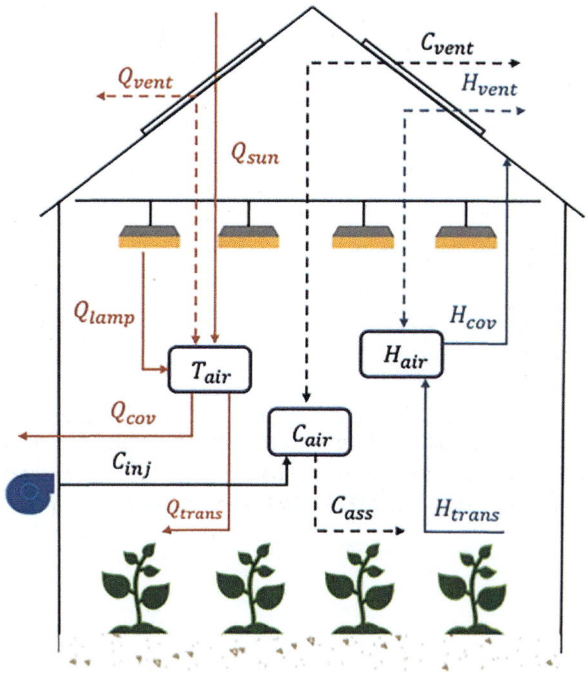

FIGURE 5.5

Overview of the greenhouse climate variables.

Van Beveren, P.J.M., Bontsema, J., Van Straten, G., Van Henten, E.J., 2015. Minimal heating and cooling in a modern rose greenhouse. Applied Energy, 137, 97–109. https://doi.org/10.1016/j.apenergy.2014.09.083.

1. Energy Savings: The model significantly reduced the energy required for heating and cooling without affecting the greenhouse climate adversely.
2. CO_2 management: Optimal use of CO_2 injection and reduced ventilation resulted in lower CO_2 consumption.
3. Grower adaptability: The method allows growers to define climate bounds, ensuring they maintain control over crop conditions while achieving energy efficiency.

Fig. 5.6 illustrates the optimal energy inputs required for maintaining specific temperature and humidity levels in a greenhouse on two different dates: February 18, 2012, and June 16, 2012. The temperature adjustments shown include increases by 0.5°C, 2.0°C, and 3.5°C above the standard settings. All other environmental controls were kept at their usual levels. The triangle symbol indicates scenarios where standard settings were optimized to achieve these conditions.

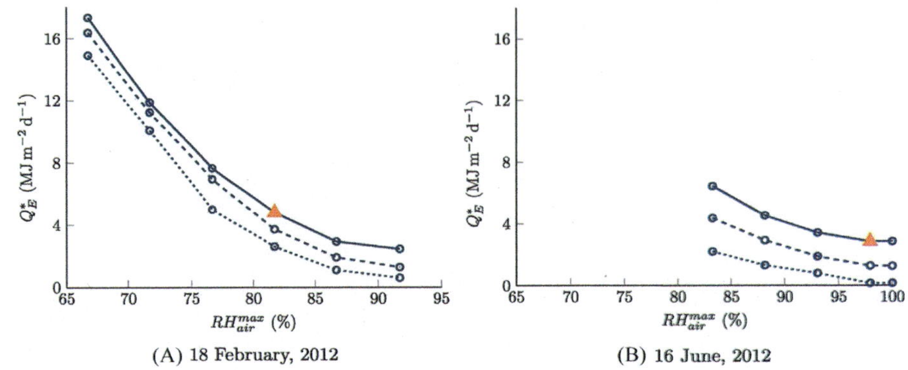

FIGURE 5.6

Optimal energy input QE_* for different temperature and humidity.

Van Beveren, P.J.M., Bontsema, J., Van Straten, G., Van Henten, E.J., 2015. Minimal heating and cooling in a modern rose greenhouse. Applied Energy, 137, 97–109. https://doi.org/10.1016/j.apenergy.2014.09.083.

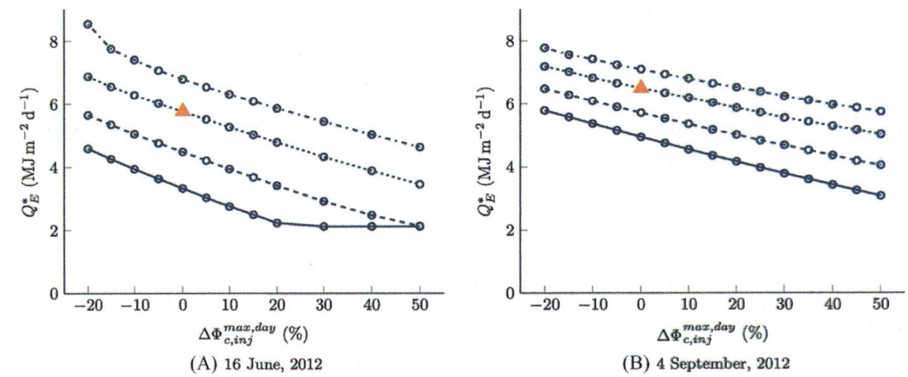

FIGURE 5.7

Optimal energy input QE_* for different lower CO_2 bounds and available CO_2.

Van Beveren, P.J.M., Bontsema, J., Van Straten, G., Van Henten, E.J., 2015. Minimal heating and cooling in a modern rose greenhouse. Applied Energy, 137, 97–109. https://doi.org/10.1016/j.apenergy.2014.09.083.

Fig. 5.7 displays the optimal energy inputs calculated for maintaining various lower CO_2 levels in the greenhouse on June 16, 2012, and September 4, 2012. The CO_2 levels considered were 20%, 13%, 3% below the standard, and an increase of

Table 5.2 Overview of total heating, cooling, and CO_2 injection settings used by the grower.

Parameter	Grower's operation	Optimal control (standard)	Optimal control (minimum pipe temperature)
Heating (GJ/ m^2/year)	2.08	1.10	1.49
Cooling (GJ/ m^2/year)	0.71	0.60	0.71
CO2 Injection (kg/m^2/year)	95.4	85.7	85.9

No permission required.

7% above the standard. These calculations assume all other environmental controls are set to their standard optimization levels. The triangle symbol represents the scenarios optimized with these standard settings.

Implementing this optimization framework can help greenhouse operators achieve substantial energy savings. The approach is adaptable and can be integrated into existing climate control systems, providing growers with actionable insights to improve energy management.

Table 5.2 compares the standard optimal settings with those where the minimum pipe temperature is maintained, illustrating different operational scenarios. The study demonstrates that optimal control techniques can lead to significant energy savings in greenhouse operations. By minimizing heating and cooling requirements while maintaining optimal growing conditions, the framework supports the goals of reducing fossil fuel use and CO_2 emissions in horticultural practices.

References

Bergland, H., Burlakov, E., Pedersen, P.A., Wyller, J., 2020. Aquaculture, pollution and fishery - dynamics of marine industrial interactions. Ecological Complexity 43, 100853. Available at https://doi.org/10.1016/j.ecocom.2020.100853.

Blasco, X., Martínez, M., Herrero, J.M., Ramos, C., Sanchis, J., 2007. Model-based predictive control of greenhouse climate for reducing energy and water consumption. Computers and Electronics in Agriculture 55 (1), 49–70. Available at https://doi.org/10.1016/j.compag.2006.12.001.

Cunha, J.B., 2003. Greenhouse climate models: an overview. EFITA 2003 Conference, 5-9. July 2003, Debrecen, Hungary. . Available at https://www.researchgate.net/publication/2930446_Greenhouse_Climate_Models_An_Overview.

Deng, Shunhua, Gao, Lihong, Zheng, Jishu, Long, Hanwei, Li, Peiyuan, 2020. Agricultural Engineering 10 (4), 41–45.

Escamilla-García, A., Soto-Zarazúa, G.M., Toledano-Ayala, M., Rivas-Araiza, E., Gastélum-Barrios, A., 2020. Applications of artificial neural networks in greenhouse technology and overview for smart agriculture development. Applied Sciences (Switzerland) 10 (11). Available at https://doi.org/10.3390/app10113835.

Iddio, E., Wang, L., Thomas, Y., McMorrow, G., Denzer, A., 2020. Energy efficient operation and modeling for greenhouses: a literature review. Renewable and Sustainable Energy Reviews 117, 109480. Available at https://doi.org/10.1016/j.rser.2019.109480.

López-Cruz, I.L., Fitz-Rodríguez, E., Salazar-Moreno, R., Rojano-Aguilar, A., Kacira, M., 2018. Development and analysis of dynamical mathematical models of greenhouse climate: a review. European Journal of Horticultural Science 83 (5), 269–279. Available at https://doi.org/10.17660/eJHS.2018/83.5.1.

Mayne, D.Q., Rawlings, J.B., Rao, C.V., Scokaert, P.O.M., 2000. Constrained model predictive control: stability and optimality. Automatica 36 (6), 789–814. Available at https://doi.org/10.1016/s0005-1098(99)00214-9.

Mehdizadeh, S., 2018. Assessing the potential of data-driven models for estimation of long-term monthly temperatures. Computers and Electronics in Agriculture 144, 114–125. Available at https://doi.org/10.1016/j.compag.2017.11.038.

Menon, P.C., 2020. IoT enabled Aquaponics with wireless sensor smart monitoring. 2020 Fourth International Conference on I-SMAC (IoT in Social, Mobile, Analytics and Cloud) (I-SMAC), Palladam, India. IEEE, pp. 171–176. Available at https://doi.org/10.1109/I-SMAC49090.2020.9243368.

Mohammadi, B., Mehdizadeh, S., Ahmadi, F., Lien, N.T.T., Linh, N.T.T., Pham, Q.B., 2021. Developing hybrid time series and artificial intelligence models for estimating air temperatures. Stochastic Environmental Research and Risk Assessment 35 (6), 1189–1204. Available at https://doi.org/10.1007/s00477-020-01898-7.

Pedersen, S., Wik, T., 2020. A comparison of topologies in recirculating aquaculture systems using simulation and optimization. Aquacultural Engineering 89, 102059. Available at https://doi.org/10.1016/j.aquaeng.2020.102059.

Piñón, S., Camacho, E.F., Kuchen, B., Peña, M., 2005. Spain Constrained predictive control of a greenhouse. Computers and Electronics in Agriculture 49 (3), 317–329. Available at https://doi.org/10.1016/j.compag.2005.08.007.

Rawlings, J.B., Muske, K.R., 1993. The stability of constrained receding horizon control. IEEE Transactions on Automatic Control 38 (10), 1512–1516. Available at https://doi.org/10.1109/9.241565.

Sethi, V.P., Sumathy, K., Lee, C., Pal, D.S., 2013. Thermal modeling aspects of solar greenhouse microclimate control: a review on heating technologies. Solar Energy 96, 56–82. Available at https://doi.org/10.1016/j.solener.2013.06.034.

Sun, J., Zhang, L., 2018. Research on fish-vegetable symbiotic environment monitoring system based on internet of things. Internet of Things Technology 8, 22–23.

Tolentino, L.K.S., Fernandez, E.O., Jorda, R.L., Amora, S.N.D., Bartolata, D.K.T., Sarucam, J.R.V., et al., 2019. Development of an IoT-based aquaponics monitoring and correction system with temperature-controlled greenhouse. Proceedings - 2019 International SoC Design Conference, ISOCC 2019 261–262. Available at https://doi.org/10.1109/ISOCC47750.2019.9027722.

Van Beveren, P.J.M., Bontsema, J., Van Straten, G., Van Henten, E.J., 2015. Minimal heating and cooling in a modern rose greenhouse. Applied Energy 137, 97–109. Available at https://doi.org/10.1016/j.apenergy.2014.09.083.

wei, g. Gao, ti lin, d he, Journal of the College of Physicians and Surgeons Pakistan. (2020), 41

Witzel, O., Wilm, S., Karimanzira, D., Baganz, D., 2019. Controlling and regulation of integrated aquaponic production systems – an approach for a management execution system (MES). Information Processing in Agriculture 6 (3), 326–334. Available at https://doi.org/10.1016/j.inpa.2019.03.007.

Yang, Y., 2020. Research progress of aquaponics systems based on Internet of Things technology. Agricultural Science-Technology and Information. 75–79.

Yep, B., Zheng, Y., 2019. Aquaponic trends and challenges – a review. Journal of Cleaner Production 228, 1586–1599. Available at https://doi.org/10.1016/j.jclepro.2019.04.290.

Zhang, S., Guo, Y., Li, S., Ke, Z., Zhao, H., Yang, J., et al., 2022. Investigation on environment monitoring system for a combination of hydroponics and aquaculture in greenhouse. Information Processing in Agriculture 9 (1), 123–134. Available at https://doi.org/10.1016/j.inpa.2021.06.006.

CHAPTER

Fish precision feeding system

6

Daoliang Li[1,2], GuangXu Wang[1,2] and Zhuangzhuang Du[1,3]
[1]*National Innovation Center for Digital Fishery, China Agricultural University, Beijing, P.R. China,*
[2]*College of Information and Electrical Engineering,China Agricultural University, Beijing, P.R. China,*
[3]*School of Computer and Information Engineering, Henan University, Kaifeng, P.R. China*

6.1 Introduction

The feeding and nutritional system of fish plays a pivotal role in the advanced agricultural paradigm of smart aquaponics. In addition to influencing the growth, health, and yield of fish, this system also affects the long-term viability of the entire ecosystem (Taha et al., 2022). Therefore, a detailed study of the operational framework and technological infrastructure of intelligent aquaponics requires dedicated research into the feeding and nutritional system.

Fish, as aquatic organisms, require appropriate nutritional supply for optimal growth and development. In smart aquaponics system, precise control over the amount, frequency, and type of feed ensures that fish receive balanced nutritional support, thereby promoting rapid and healthy growth (Yanes et al., 2020). This precision enhances farming efficiency, reduces costs, and minimizes the incidence of fish diseases caused by nutritional deficiencies or excesses. Furthermore, the fish feeding and nutrition system is integral to achieving ecological cycling and sustainable development within the smart aquaponics framework. In this system, fish waste serves as organic fertilizer for plants, promoting their growth, while plants produce oxygen through photosynthesis, creating essential conditions for fish survival. This mutually beneficial relationship forms a closed-loop ecosystem, maximizing resource utilization and minimizing waste emissions.

The optimization and innovation of the fish feeding and nutritional system in intelligent aquaponics serves to enhance the efficiency of farming operations, while simultaneously driving the transformation and upgrading of the entire agricultural industry (Xu et al., 2024). The achievement of such a virtuous cycle necessitates the implementation of scientific planning and meticulous management of the fish feeding and nutrition system. This necessitates the formulation of appropriate feed recipes based on the growth stages and nutritional needs of the fish, the adjustment of feeding amounts and frequencies according to environmental factors such as water quality, temperature, and fish biomass, and the employment of modern technologies for real-time monitoring and intelligent control of the feeding process.

Smart Aquaponic Farming. DOI: https://doi.org/10.1016/B978-0-443-28945-3.00007-4

The integration of technologies such as the Internet of Things (IoT), big data, and artificial intelligence (AI) enables the comprehensive perception and precise control of the fish growing environment, thereby enhancing the automation and intelligence of aquaculture. Furthermore, the optimization of feed formulations and feeding strategies can reduce feed costs, energy consumption, and environmental pollution in the aquaculture process.

This chapter presents an analysis of the theoretical and practical aspects of fish feeding and nutritional systems in intelligent aquaponics. It discusses the calculation of fish biomass, the development of feeding strategies, the design of feeding control systems, the concept of feed conversion, and the estimation of waste. Furthermore, this chapter examines the potential of modern technologies for optimising the fish feeding and nutritional system, with the objective of enhancing both aquaculture efficiency and sustainability.

6.2 Fish biomass calculation

The estimation of fish biomass in smart aquaponics systems is of paramount importance to aquaculture managers, as it allows for optimization of feeding rates, regulation of stocking densities, and determination of optimal harvest times. The traditional methodology for measuring fish weight entails periodically capturing a sample of fish from cages or ponds, administering an anesthetic, and manually weighing them (Zhang et al., 2020). This approach is not only labor-intensive and time-consuming but also induces stress or injury to the fish, which may result in mortality in severe cases. Consequently, there is an urgent need for non-invasive and automated methods for measuring fish weight. The advent of machine vision technology has led to the development of a rapid, accurate, and non-contact method for measuring fish size (Hao et al., 2022). Given that fish body size is closely correlated with their weight, it is now possible to estimate fish biomass automatically based on fish size without the need to remove them from the water. This approach has the potential to significantly reduce stress on the fish and enhance their welfare. The machine vision system used for biomass estimation is shown in Fig. 6.1.

6.2.1 Fish size measuring system

In smart aquaponics systems, the real-time monitoring and assessment of fish growth are of critical importance in order to ensure the efficient operation of the system. The size of the fish, which is a key indicator of their growth status, directly influences the farming strategies employed and the yield predictions (Shi et al., 2020). Conventional direct measurement techniques, which employ instruments such as rulers or calipers to ascertain the dimensions of captured fish, are relatively simple and cost-effective. However, these methods may induce stress in fish due to handling and capture, which could potentially affect their growth. Furthermore,

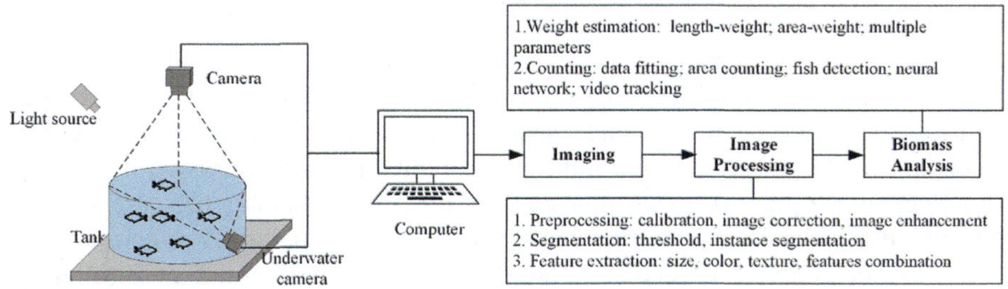

FIGURE 6.1

The machine vision system for biomass estimation.

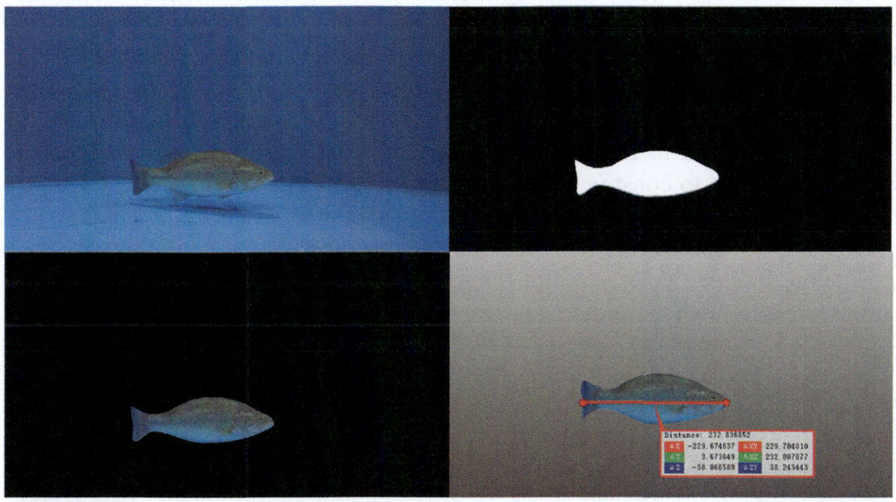

FIGURE 6.2

The measurement of fish size of based on machine vision.

these methods are labor-intensive, inefficient for large-scale farming, and may lead to fish injuries or mortality, which in turn impacts overall yield (Li et al., 2020). Consequently, direct measurement is more appropriate for use in small-scale experimental studies or in cases where only occasional sampling is required.

Machine vision techniques are frequently employed in smart aquaponics for the measurement of fish size, due to non-invasive, high automation, and real-time capabilities. The measurement method of fish size based on machine vision is shown in Fig. 6.2. This approach entails the capture of images of fish within the tank by means of high-definition cameras, subsequent application of image recognition technology for the extraction of fish contours and the calculation of size (Zhao et al., 2021). This method is conducive to minimal stress upon the fish, as it does not necessitate their removal from the water, and it permits the implementation of automated, real-time monitoring. Cameras can be mounted above the fish tanks, capturing images at regular intervals or when triggered. The images are then analyzed by image recognition algorithms, which identify key features such as the head, tail, and various body parts. This enables the accurate calculation of the fish's length, width, and volume. The non-intrusive nature and high level of automation make this method ideal for reducing fish stress while collecting extensive data for further analysis and decision-making. Furthermore, the system can be adapted to accommodate different fish species and sizes, aligning with the diverse needs of aquaculture. As computer vision and artificial intelligence technologies evolve, the precision and efficacy of image recognition techniques are anticipated to enhance (Saberioon et al., 2017).

Three-dimensional (3D) scanning methods create a detailed model of the fish, providing precise measurements. These methods employ laser scanners or structured light scanners, which capture comprehensive 3D information, including length, width, height, and volume. By scanning the fish from multiple angles, high-resolution 3D point cloud data are generated. The 3D reconstruction results of the fish body are shown in Fig. 6.3. These data are then processed and analyzed by computer software to construct a 3D model of the fish and extract the required size information. While 3D scanning offers high precision and data completeness, it is costly, complex, and has stringent environmental requirements, limiting its practical application. It is particularly useful for research settings demanding high accuracy or for managing high-value fish species.

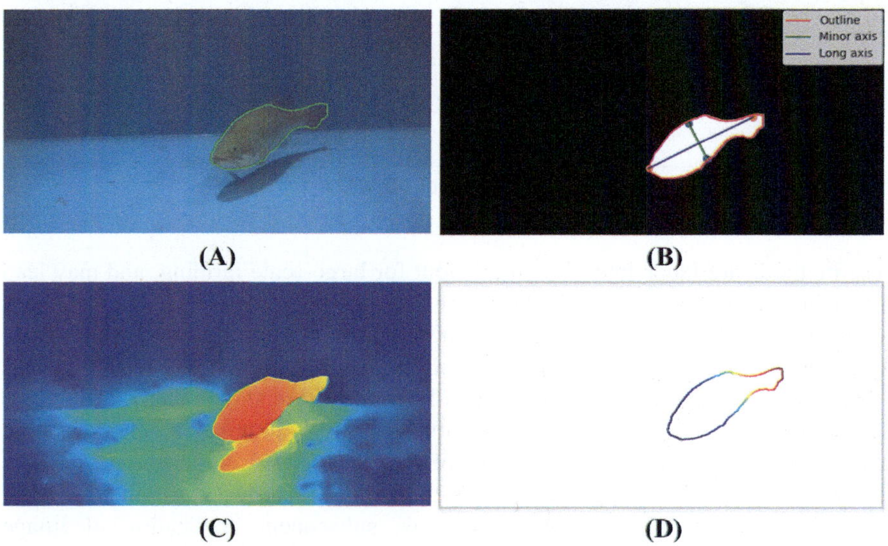

FIGURE 6.3

The 3D reconstruction results of the fish contour.

The selection of an appropriate fish size measurement method is of paramount importance in the field of smart aquaponics. Direct measurement methods are suitable for small-scale experiments and necessary sampling, while image recognition methods are widely used due to their non-invasiveness and high degree of automation. Three-dimensional scanning has the potential to be a valuable tool in specific scenarios where high precision is required. Fish size measurement systems are applicable not only to fish farming management in smart aquaponics, but also to a range of other aquaculture domains. As a result of ongoing advancements in artificial intelligence, these systems will play a vital role in enhancing the efficiency of farming operations, reducing the intensity of labor required, and promoting sustainable development in the aquaculture industry.

6.2.2 Fish weight estimation system

Accurately estimating fish weight in smart aquaponics systems is crucial for monitoring growth, formulating farming strategies, and predicting biomass. Image processing techniques have emerged as a focal point of research, with primary methods including length-based estimation, area-based estimation, and multi-parameter estimation. The architecture of the fish weight estimation system is shown in Fig. 6.4.

In monocular vision systems, a single camera is utilized to capture images of fish. Subsequently, image processing techniques are employed to extract the length of the fish, which is then used in a length–weight relationship model to estimate the weight (Gümüş et al., 2021). This method is advantageous due to its low equipment cost and simplicity, rendering it suitable for fish that are either stationary or moving slowly. However, the presence of complex backgrounds and variations in fish posture can result in a reduction in measurement accuracy. In contrast, stereo vision systems employ two or more cameras to capture images of fish from different

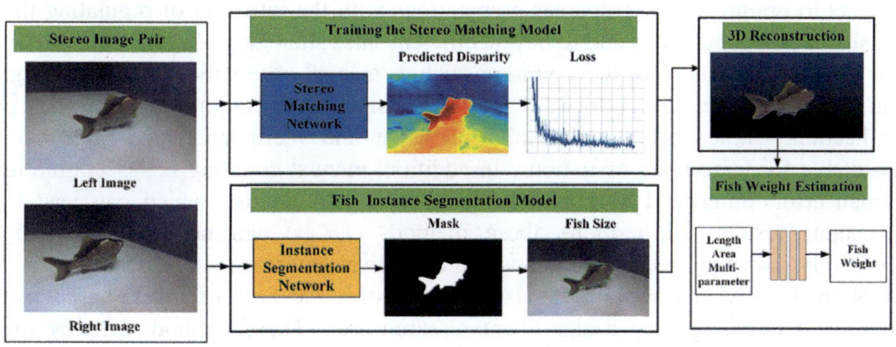

FIGURE 6.4

The architecture of the fish weight estimation system.

angles, thereby calculating three-dimensional information for the precise measurement of fish length and weight (Zhang et al., 2024). Although this method offers high measurement accuracy and is suitable for dynamically swimming fish, it is more complex and costly, with high demands on data processing.

Area-based weight estimation methods have demonstrated high accuracy and reliability, making them a significant approach for estimating fish weight. The body area of fish, which is a parameter independent of orientation or curvature, is a commonly used parameter in weight estimation. Through image processing techniques, the number of pixels representing the fish is converted into the fish's area, and a relationship model between the area and weight is established. However, research in this area has primarily focused on post-harvest weight estimation. The development of non-invasive automatic algorithms for the removal of tail fins and the validation of their impact on the accuracy of mass models will contribute to the precise measurement of fish mass during swimming (Hao et al., 2022).

To improve the accuracy of fish weight estimation, some researchers have integrated multiple parameters, including area, length, height, and perimeter. It is recommended that future studies focus on improving the accuracy of image processing algorithms for underwater images, enhancing the precision of three-dimensional size measurement and weight estimation of swimming fish and exploring the potential of integrating deep learning and machine learning techniques. These developments will further enhance the automation and reliability of the system. The ongoing advancement of these technologies will propel the evolution of smart aquaponics towards enhanced precision and efficiency, thereby facilitating sustainable development in aquaculture.

6.2.3 Fish counting system

In smart aquaponics, the estimation of the number of fish at different growth stages is of paramount importance for the effective management of the system. This enables farmers to optimize their container preparations with the intention of regulating the density of fish populations and to devise efficient sales strategies (Zhang et al., 2020). The traditional methods of fish counting, such as manual observation, hand-catching, and weighing, are straightforward but have significant limitations in practical application. These methods are labor-intensive and inefficient, rendering them unsuitable for large-scale operations. In addition, manual counting is susceptible to human error, influenced by subjective judgement and fatigue, which can lead to inaccurate results. Furthermore, these methods can be detrimental to the fish, negatively impacting their growth and the ecological environment.

With the advent of computer vision technology, the field of image-based fish counting has emerged as a significant research area. These methods employ the capture of images of fish in ponds or tanks, with subsequent image processing and analysis techniques employed to automatically count the fish. The architecture of the fish counting system is shown in Fig. 6.5. In the context of static image-based

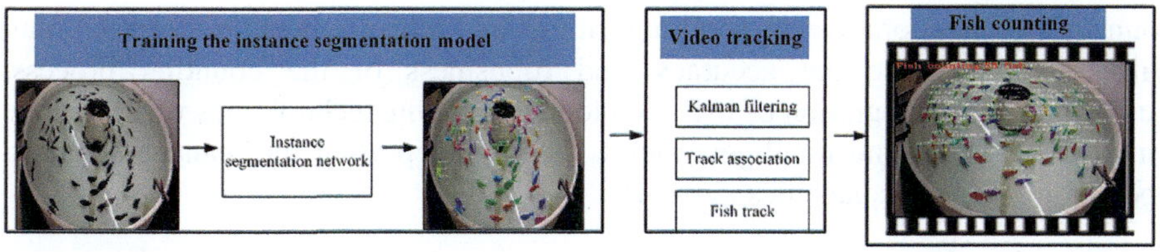

FIGURE 6.5

The architecture of fish counting system.

counting, high-definition cameras are employed to capture still images of the fish habitat. Techniques such as image segmentation, feature extraction, and classification permit the automatic counting of fish (Liu et al., 2021). These methods are non-invasive and provide real-time data, although they do have limitations in the case of overlapping or occluded fish. Video-based counting methods employ the use of continuous video images, which are analyzed using techniques such as object tracking and motion detection in order to provide real-time dynamic fish counts (Li et al., 2020). These methods provide a more accurate reflection of fish population changes, but require substantial computational resources and high-performance hardware.

The application of machine learning and deep learning techniques is pervasive in the domain of fish counting. The application of machine learning or deep learning algorithms during the training process enables the automatic recognition and counting of fish images. For instance, random forest algorithms classify fish images by integrating multiple decision tree models, thereby enhancing both accuracy and robustness in the handling of complex image information and nonlinear relationships (Zion, 2012). Support vector machines are also employed for feature extraction and classification. The algorithm performs well with small samples and high-dimensional features, but it is less effective when the fish are overlapping or occluded (Zhou et al., 2017). Recent advances have included the utilization of convolutional neural networks and recurrent neural networks for the purpose of feature learning and classification of fish images (Höfener et al., 2019). These deep learning models are capable of handling complex image data with high accuracy and robustness; however, they require extensive labeled data for training.

In addition to image-based and machine learning methods, object tracking-based fish counting systems represent a current area of research focus. These systems employ tracking algorithms to monitor fish movement trajectories, thereby providing real-time, dynamic counts. Object tracking algorithms, including Kalman filters, particle filters, and deep learning, offer accurate reflections of fish population changes and can perform multi-object tracking and identification. As a consequence of the unceasing technological progress and innovation, future fish counting methods and systems will become increasingly intelligent, efficient, and precise. The integration of image synthesis techniques, which merge data from multiple cameras, can provide comprehensive perspectives and more accurate counts. The incorporation of novel algorithms and advancements in deep learning will further enhance the accuracy and robustness of the counting process. Furthermore, the integration of IoT and cloud computing technologies will facilitate remote monitoring and intelligent alerts, thereby supporting the management and operation of smart aquaponics systems.

6.3 Fish feeding model

Feeding in real production mainly focuses on manual judgment and mechanical quantitative feeding, which cannot be fed according to the actual demand of the fish and easily lead to overfeeding or underfeeding. Recently, the construction of feeding models has been regarded as a key step in realizing self-demand feeding, and the study of feeding intensity is an important part of the construction of feeding models. Fish feeding intensity recognition models can be categorized into three types, including water quality sensor-based, machine vision-based, and sound-based.

Specifically, changes in water quality (e.g., water temperature, dissolved oxygen concentration, pH, ammonia-nitrogen compounds, etc.) can directly affect fish appetites. Temperature can affect the metabolic rate of fish and, thus, their ability and desire to acquire food. High dissolved oxygen promotes the feeding intensity of fish. Nevertheless, changes in the concentration of ammonia and nitrogen compounds caused by localized reductions in dissolved oxygen concentration due to fish fighting and the deposition of unfed pellets on the bottom during the feeding period are not evident and only reflect the feeding status of the fish to a certain level.

Compared to assessing the feeding intensity of fish through changes in water quality parameters, machine vision technology is more intuitive, convenient, fast, and highly accurate (Yang et al., 2021; Zhou et al., 2019). The 3D ResNet-Glore fish feeding intensity quantification network proposed by Feng et al. can accurately localize four levels of fish feeding behavior in video streams (Feng et al., 2022). Wang et al. constructed a multitasking network for continuous monitoring of fish feeding activities and the number of uneaten feed pellets in the fish tank, and the results showed that the multitasking network achieved 95.44% accuracy in classifying fish feeding behaviors (Wang et al., 2022). Although vision-based methods have made some progress in assessing the intensity of fish feeding, their real application in factory aquaculture is still limited by the dim environment, turbid water, reflective area texture, environmental complexity, and variability, which makes it difficult to achieve accurate real-time monitoring and prediction.

Recently, besides assessing the feeding intensity of fish by water quality parameters and visual techniques, the production of pulsed acoustic signals by fish and shrimp during feeding can be used as another important indicator of feeding activity (Silva et al., 2019). Wei et al. established the corresponding amplitude relationship between the characteristic parameters of the vocalizations of the shrimp *Penaeus vannamei* and the degree of hunger based on passive acoustic techniques as a basis for determining on-demand feeding (Wei et al., 2020). Zeng et al. (2023) proposed an audio-spectral model based on acoustic signals and attentional mechanisms that classified the feeding intensity of fish into four classes. Du et al. (2023) proposed a feature fusion-based feeding intensity recognition scheme that fused Mel Spectrogram, STFT, and CQT feature maps, resulting in an accuracy of 97.941%. The use of passive acoustic methods of detection does not negatively affect the feeding environment and feeding behavior of fish and shrimp and provides a basis for developing more effective feeding strategies to match feeding regimes to the feeding needs of different fish populations.

From the aforementioned analysis, it can be seen that single-modality recognition tasks have been widely studied in recent years, and some remarkable results have been achieved. However, since single-modality methods observe only part of the information about the thing of interest, which is susceptible to single-modality noise, fusion of multimodal information based on its consistency to improve the robustness of the model has attracted attention over the years.

Multimodal fusion integrates information from several different data sources to obtain more comprehensive and accurate information to support richer analytics, decision-making, and applications. These data sources include images, video, sound, text, sensor data, and others. In general, multimodal information fusion strategies can be categorized into early, intermediate, and late stages (Baltrusaitis et al., 2019). Early fusion constructs a joint representation at the input level based on raw data or features, which are then fed into the model. Intermediate fusion enables combining individual modalities at different levels of abstraction. In particular, within a system with three or more modalities, data may be fused at once or progressively at different levels. Late fusion trains a single model for each modality and aggregates predictions from a single model at the decision layer. Recently, multimodal fusion has been playing an important role in many fields, such as medical image analysis (Steyaert et al., 2023), speech recognition and natural language processing (Chaudhari et al., 2023), intelligent transportation systems, and agricultural remote sensing (Chen et al., 2017).

The fish feeding intensity model presented in this book is based on the multimodal fusion technique, which integrates water quality parameters, visual information, and biomass to judge fish feeding. We would like to focus on this model. Specifically, the model consists of three modules, namely, the water quality parameter processing module in Section 6.3.1, the biomass calculation module in Section 6.3.2, and the machine vision evaluation module in Section 6.3.3. Finally, the modal information of each module was integrated to create a multimodal model to overcome the single-modal difficulty in fully reflecting the one-sidedness of the feeding state, to improve the accuracy of the quantification of feeding intensity, and to provide technical support for the development of a precision feeding system.

6.3.1 Water quality parameter processing module

Critical water quality parameters such as pH, dissolved oxygen, and temperature were acquired by multi-parameter water quality sensors. The acquisition interval of water

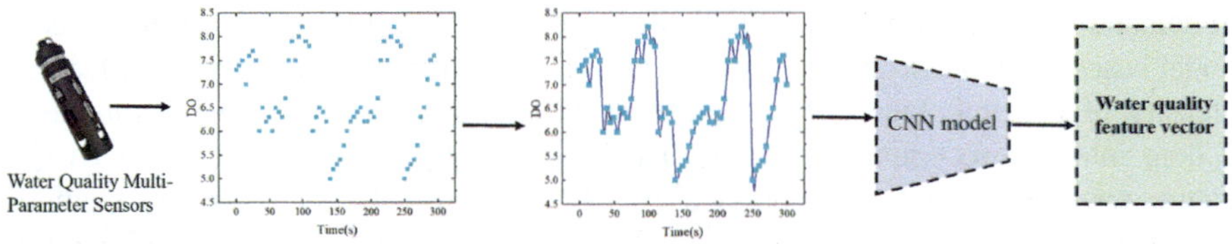

FIGURE 6.6

Water quality feature extraction.

FIGURE 6.7

Hikvision video feature extraction process.

quality data is 5s, and water quality data processing is shown in Fig. 6.6. First, the water quality data of the whole feeding phase were fitted to a function, and the corresponding video time periods were sampled along the fitted curve at intervals of 0.01s. The pH, dissolved oxygen, and temperature data were obtained and normalized. Eventually, water quality features such as pH, dissolved oxygen, and temperature are concatenated to obtain the feature vectors of the water quality parameters.

6.3.2 Biomass calculation module

Image recognition methods are non-invasive, highly automated, and real-time and are commonly used to measure fish size in smart aquaponics (Hao et al., 2022). This method involves using a high-resolution video camera to capture images of fish in a tank and applying image recognition techniques to extract the outlines of the fish and calculate their size. This method minimizes stress on the fish as it does not need to be removed from the water and allows for automated real-time monitoring. Please refer to Section 6.2 for a detailed description of this part.

6.3.3 Machine vision evaluation module

The process of feature extraction based on the video data obtained from Hikvision cameras is shown in Fig. 6.7. A data set of 2s or 3s can be constructed. Since the frequency of the original video sampling is 50 Hz and the resolution of each frame is 2560 × 1440, there is a large amount of redundant information in the multi-frame video. In order to solve this problem and reduce the computational complexity, a downsampling technique can be adopted. Specifically, a frame is randomly selected from the video, and then each frame is resized to a smaller size of $n \times n$ (where n is usually taken to be 224). This resize step aims to preserve the essential visual information while making the data more easily manageable for training purposes. By combining downsampling, random cropping, and color enhancement techniques during the training process, the high-resolution video data can be effectively utilized while improving the model's ability to generalize to different inputs. The video frames of size $n \times n$ are then fed into the CNN network model to extract the video feature vectors, respectively.

6.3.4 Multimodal fusion of fish feeding models

The framework of multimodal fusion of fish feeding models (MFFF) is shown in Fig. 6.8. The deep-level features of water quality were obtained by the water quality parameter processing module, and then, after a classifier, the fish feeding prediction result P_T based on water quality parameters was obtained. Meanwhile, the video of the fish feeding process was obtained by the machine vision evaluation module. Deep-level features of the video during fish feeding were obtained by the machine vision evaluation module, and then, after a classifier, the vision-based fish feeding prediction result P_V was obtained. In addition, the biomass-based fish feeding prediction result (P_B) was also obtained by the method in Section 6.2.

6.3.5 Summary

The multimodal fusion approach can focus on more information about the fish feeding process compared to a single modality (audio, visual, and acoustic). This characteristic point demonstrates that this approach is sufficient to capture the heterogeneity of the fish feeding process and can better achieve fish feeding intensity classification. The successful integration of text, visual, and other data is

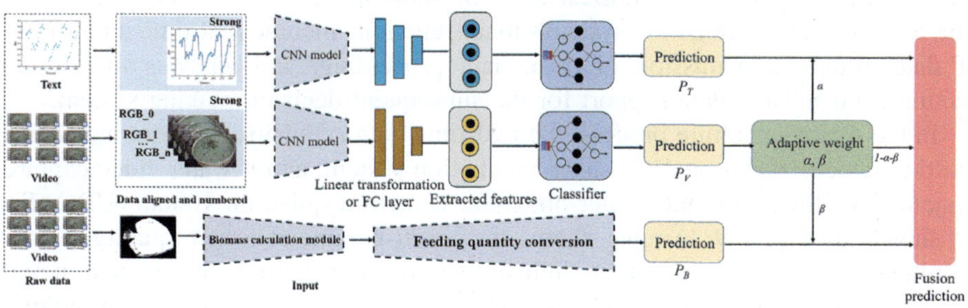

FIGURE 6.8

Multimodal fusion of fish feeding intensity frameworks.

expected to further improve our understanding of fish feeding and allow for the development of precise and personalized feeding strategies, which provides the feasibility of realizing a better application of quantification of fish feeding intensity on the mobile device.

6.4 Feeding system

The "Fish Feeding System" consists of four main modules: The first is the real-time data acquisition system, which is responsible for collecting data on fish growth, water quality parameters, and others; The next is a cloud platform decision-making system that analyzes the data and combines it with pre-programmed algorithms to provide feed delivery suggestions; then there is the feeding control system, which precisely controls the time and amount of feeding according to the instructions from the cloud platform system. Finally, there is the rail-mounted automatic feeding system, which is responsible for realizing automatic feed placement according to the instructions of the control system in accordance with the rail. The overall composition achieved the functions of data collection, analysis, decision-making, and automated feeding, which improved breeding efficiency and cost control.

6.4.1 Real-time data acquisition system

Real-time data acquisition systems normally consist of several key components, which include sensor networks, data transmission modules, data processing units, and memory devices. The sensor network is responsible for real-time monitoring of fish growth, water quality parameters, and other key data such as temperature, dissolved oxygen, and pH. The data transmission module transmits the data gathered by the sensors to the data processing unit. The data processing unit is responsible for parsing, processing and analyzing the sensor data to generate data that can be available to the cloud platform system; Storage devices are used for the long-term preservation of historical data for subsequent analysis and comparison. The real-time data acquisition system's main functions include real-time monitoring of data, timely transmission of data, data processing, and analysis to provide accurate and reliable data support for the subsequent decision-making system.

In the real aquaculture process, the real-time data acquisition system includes a multi-parameter water quality sensor acquisition system, an on-water video capture system (VCS), and an underwater binocular camera acquisition system (UBCAS), as shown in Fig. 6.9. Among them, the multi-parameter water quality sensor monitors a variety of water quality indicators in the aquaculture environment in real time, such as temperature, dissolved oxygen, pH, and other parameters, providing instant and accurate data. Further, the collected data is transmitted to the data collector via wires, and then the data is sent to the local server, usually by means of Wi-Fi, Ethernet, 4G/5G, LoRaWAN, and others. Then a suitable data transfer

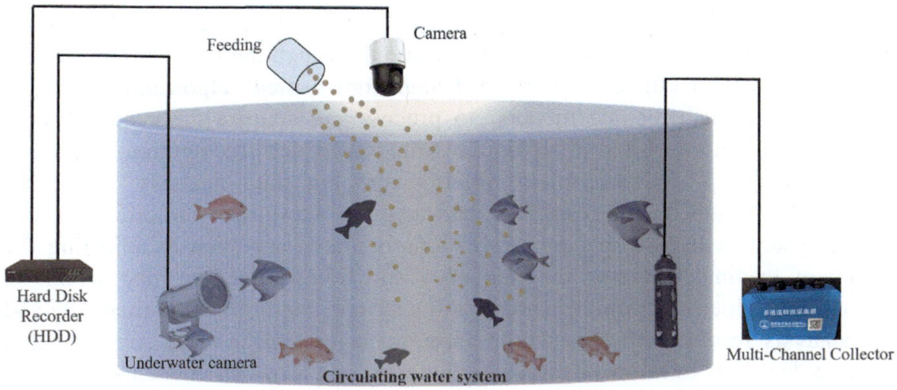

FIGURE 6.9

Real-time data acquisition system.

protocol, such as MQTT (Message Queuing Telemetry Transport), HTTP, CoAP (Constrained Application Protocol), and so on, is chosen for transferring the locally collected data to the cloud platform. VCS and UBCAS, on the other hand, require configuration of the network, selection of encoding formats, transmission protocols, and the writing of programs to ensure real-time transmission to the cloud and monitoring of the data.

6.4.2 Cloud platform decision-making system

The cloud platform decision-making system in the "fish feeding system" usually consists of a data-receiving module, a data-processing module, a decision-algorithm module, and a user-interface module. The data receiving module receives data from the real-time data acquisition system, including fish growth, water quality parameters, and so on; the data processing module is responsible for cleaning, pre-processing, and integrating the received data. The decision-making algorithm module integrates pre-defined algorithms and models to analyze the processed data and provide decision-making suggestions regarding feeding time and amount; the user interface module provides a visual interface that enables the user to view the data, receive decision-making suggestions, and perform the necessary manual interventions. Based on real-time data and algorithmic models, the main function of the cloud platform decision-making system is to provide scientific and reasonable feeding suggestions in order to realize intelligent management and optimization of the breeding process.

During the real aquaculture process, real-time data acquired by multi-parameter water quality sensors, an above-water video capture system, and an UBCAS from aquaculture farms, including water quality, water temperature, dissolved oxygen,

PH value, and video of the fish feeding process, is transmitted to the cloud platform, and real-time data updating and storage is ensured through the network connection. The cloud platform utilizes a developed algorithm named "algorithm for multi-modal fusion of fish feeding models" to process and analyze the acquired data; meanwhile, it can also be used for data mining and pattern recognition of feeding process intensity through artificial intelligence, big data analysis, and other technologies. Based on the results of data analysis, the cloud platform decision-making system will give specific suggestions or decisions, such as adjusting the amount of feeding, changing the frequency and time of feeding, and so on. In summary, the cloud platform decision-making system can utilize real-time data and data analysis results, which can help farm managers make more scientific and effective decisions to improve farming efficiency and economic benefits.

6.4.3 Feeding control system

The "feeding control system" is a very critical component of the' fish feeding system," whose structure includes a controller, actuators, and sensors. The controller receives commands from the decision-making system of the cloud platform and controls the time and amount of feeding according to the set parameters; the actuator is responsible for the actual execution of feeding operations, such as opening the feeding port or controlling the placement of the feeder; the sensors are used to monitor the indicators during the feeding process, such as the response of the fish population, feed consumption, and so on, to feed back data to the controller for real-time adjustment. The primary function of the feeding control system is to accurately control the time and amount of feed delivery according to the decision-making suggestions from the cloud platform system to ensure that the fish get the amount of feed they need to improve farming efficiency and reduce waste. After integrating the feeding control system, the aquaculture process is more intelligent and precise.

6.4.4 Rail-mounted automatic feeding system

The "Rail-mounted automatic feeding system" is a further development on the basis of a feeding control system that utilizes automation technology, precise positioning, and automatic feed placement by using a rail system to achieve more efficient and precise feeding management, which improves breeding efficiency and reduces cost. The rail-mounted automatic feeding system is an efficient automatic feeding device for farms, the structure of which includes a rail system, a feeding device, and a control system (Fig. 6.10). The rail system usually consists of fixed rails along the aquaculture pond or water area and is equipped with motorized drives for moving the feeding device. The feeding unit is mounted on a rail system that can be moved along the rails above the aquaculture ponds and performs feeding operations according to a pre-defined path and schedule. The control system is used to monitor

FIGURE 6.10

Rail-mounted automatic feeding system.

the operating status of the feeding system and regulate the feeding time and amount, which can be remotely controlled and adjusted according to demand. The primary functions of the rail-mounted automatic feeding system include (1) Automated feeding: According to the preset feeding schedule and parameters, the system is able to automatically feed the fish in the breeding pond along the rail in a quantitative manner (Fig. 6.11). (2) Timing and quantitative feeding: Different feeding times and quantities can be set according to the cultivation demand and the growth of the fish to ensure that the fish get the right amount of feed. (3) High efficiency and saving: Through automated feeding, it can reduce the labor cost of manual feeding, improve the accuracy of feeding, and avoid waste and overfeeding. (4) Data monitoring and recording: The system is usually equipped with data collection and recording functions, which can monitor the indicators in the feeding process and generate feeding records and reports to provide data support for breeding management. The rail-mounted automatic feeding system can improve breeding efficiency, reduce labor costs, and realize automation and intelligent management of the breeding process.

6.5 Feed conversion and waste calculation system

In smart aquaponics systems, the feed conversion ratio (FCR) and waste management systems are critical components that directly impact aquaculture efficiency and economic viability, as well as resource utilization and environmental protection. The FCR is a pivotal metric for gauging the efficiency with which animals

FIGURE 6.11

Automatic feeding of fish tanks in aquaponics scenarios.

convert feed into body mass, and it is of particular significance in the context of the sustainability of smart aquaponics operations (Palm et al., 2024). An efficient FCR implies that a greater proportion of the feed is converted into biomass, which in turn reduces waste, increases yield and lowers costs. In order to obtain accurate FCR calculations, it is essential to ensure that the measurement and recording of both the feed input and the animal weight gain are precise. The implementation of advanced automated feeding and weight monitoring systems in smart aquaponics facilities enables the real-time acquisition of data and the subsequent automated calculation of relevant parameters.

The waste management system in smart aquaponics entails the quantification and management of a range of waste products generated during aquaculture, including uneaten feed, feces, and other excretions (David et al., 2022). Effective waste management is a crucial aspect of improving resource utilization and minimizing environmental pollution. It is imperative that the system be capable of accurately classifying and measuring all types of waste. Sensors and monitoring devices are employed to track key data, including feed input, consumption, and waste production. The application of image processing and analysis technologies enables the automatic identification and quantification of feed residues and feces.

In addition, the waste management system carries out a quantitative analysis of the waste (Araújo et al., 2021). By collecting and analyzing data on the composition, weight and volume of waste, the system can assess the potential for resource recovery and the environmental risks associated with the waste. This data-driven approach helps to develop treatment strategies aimed at waste reduction, resource recovery and minimizing environmental impact. Automated waste treatment and recycling processes are an integral part of the system. Technologies such as composting, biogas production and biodegradation can be used, and the integration of these technologies with automated equipment and control systems enables efficient waste collection, segregation, treatment, and recycling. This integration not only reduces operating costs and increases economic benefits, but also reduces pollution and maintains the ecological balance.

The implementation of smart aquaponics facilities can lead to improvements in fFCR and a reduction in waste generation. This can be achieved through the optimization of feed formulations, the implementation of precise feeding strategies, the enhancement of health management, the utilization of data analytics, and the promotion of resource recycling (Miličić et al., 2017). The formulation of scientific feed compositions based on the growth stages and nutritional needs of the animals ensures that they receive adequate nutrition while minimizing waste. Automated feeding systems can deliver precise amounts of feed based on real-time needs, preventing overfeeding and subsequent waste. The implementation of health management practices that prevent disease and maintain high health standards also contributes to an improvement in FCR and a reduction in waste. The use of advanced data analytics can uncover insights from FCR and waste data, identifying potential issues and optimization opportunities to refine system operations. Resource recovery initiatives, such as the use of leftover feed and feces as fertilizer, can lower costs and reduce the environmental impact.

In conclusion, the FCR and waste management systems in smart aquaponics are complex yet vital components of the overall system. The accurate calculation and effective management of FCR and waste production can enhance aquaculture efficiency, reduce costs, and protect the environment. As technology continues to advance, it is reasonable to anticipate the emergence of more efficient and precise systems, which will further support the development of smart aquaponics and contribute to the sustainable growth of the aquaculture industry.

6.6 Challenges and future prospects

Real-time fish feeding system based on multimodal fusion

6.6.1 The behavioral complexity of farming objects

Fish in a subnormal state such as scare or stress will have a weakened desire to feed or may even stop feeding altogether, at which point efforts at the technical level will be ineffective. Therefore, it is a key challenge to accurately monitor the abnormal behaviors of cultural subjects and establish the link between abnormal behaviors and feeding desires.

Outlook: Application of sensor technology: Various types of sensors, such as cameras, sonar, motion sensors, and so on, are used to monitor the behavior of farmed animals. These sensors can capture data such as activity patterns, location, and feeding of farmed organisms. **Data analysis and artificial intelligence:** monitored data are processed and analyzed using data analysis and artificial intelligence techniques. By monitoring changes in behavioral patterns, it is possible to identify abnormal behavior and establish a relationship with the feeding desire. **Establishment of a behavioral model**: Based on the monitored data, the behavioral model of the breeding object is established, including the characteristics of normal and abnormal behaviors. Abnormal behavior can be detected in time by comparing the actual observed data with the model-predicted data. **Feeding behavior analysis**: Combined with the monitored feeding data, analyze the correlation between feeding behavior and abnormal behavior. For example, observe whether the amount of food consumed by the breeding animal is related to the frequency of abnormal behaviors or whether abnormal behaviors affect the feeding desire. **Real-time monitoring and feedback:** Establishing a real-time monitoring system can detect abnormal behaviors in time and regulate the breeding environment, feed delivery, and other factors through system feedback in order to maintain the health status and feeding desire of the breeding objects. Through the integrated application of the above methods, which can effectively monitor the abnormal behavior of aquaculture objects as well as establish the connection between abnormal behavior and feeding desire, so as to realize accurate aquaculture management and regulation.

6.6.2 Complexity and mutability of the farming environment

Environmental factors such as changes in water quality, sudden drops in water temperature, and current speed can directly affect the feeding status of fish, and maintaining the systematic stability of the aquaculture environment poses a significant challenge to the development of an intelligent feeding system.

Outlook: real-time monitoring of water quality and environmental parameters: Installing water quality monitoring equipment and environmental parameter sensors, regularly monitoring water quality, water temperature, dissolved oxygen, pH value, ammonia nitrogen and other key parameters. Real-time knowledge of the state of the environment and timely detection of abnormalities. **Establishment of an automated control system:** Establishment of an automated control system to adjust environmental factors such as water quality, water temperature, and water flow rate in real time according to monitoring data to maintain a stable breeding environment. For example, automatic control of water pumps, heaters, ventilation equipment, and so on. **Set alarm mechanism:** Set the threshold value of environmental parameters, which will trigger an alarm when it exceeds the set range, and remind the breeding personnel to take timely measures. A timely response can be realized through a mobile app or alarm. **Maintain the equipment regularly:** Check and maintain the water quality monitoring equipment and automation control equipment regularly to ensure their normal operation. Timely replacement of aged and damaged equipment to avoid environmental stability decline due to equipment failure. **Emergency plan formulation:** Develop an emergency plan for water quality changes, sudden drops in water temperature, and other emergencies, and clarify the treatment process and responsible persons to ensure the stability of the breeding environment. **Clean aquaculture facilities regularly:** Clean aquaculture facilities, including fish tanks and filters, on a regular basis to keep the water quality clean, reduce bacterial growth and the accumulation of harmful substances, and maintain the stability of the aquaculture environment. Through the comprehensive application of the above measures, the systematic stability of the aquaculture environment can be effectively maintained, and aquaculture efficiency and fish growth quality can be enhanced.

6.7 Generalization capability of the model

Most of the existing models developed are specific to a particular species of fish within a particular cultural environment. Due to differences in feeding habits and other aspects of different species of fish, there is still a need to validate the performance of such methods in assessing the feeding intensity of other fish species to improve their applicability on commercial farms.

Outlook: Diversity of data collection: Collect data from different species of fish and different cultural environments, including feeding habits, growth, environmental parameters, and so on, to ensure that the data are representative and diverse. **Characterization engineering:** To design suitable characterization engineering methods for different species of fish and different cultural environments, and to extract key characteristics to reflect feeding habits and growth status under different conditions. **Model selection and tuning:** Choose models suitable for processing multimodal data, such as deep learning models, integrated learning models, etc., and perform parameter tuning on the models to improve the generalization ability. **Cross-validation:** Cross-validation methods are utilized to assess the generalization performance of the model to ensure stable performance on different datasets and to reduce the risk of overfitting. **Migration learning:** Utilizing migration learning techniques to fine-tune existing models to adapt to new varieties and environments, accelerating the ability of models to adapt in different scenarios. **Model integration:** Fusion of predictions from different models can improve the robustness and generalization ability of the models to adapt to a wider range of fish species and aquaculture environments. **Continuous learning and iterative optimization:** Continuously collecting new data and updating the model to achieve continuous learning and iterative optimization, so that the model gradually adapts to the needs of different species of fish and aquaculture environments. Through the combined application of the above strategies, the generalization ability of the model can be improved so that it can be effectively applied to different species of fish and different scenarios of aquaculture environments, providing a wider range of decision support and management optimization solutions for the aquaculture industry.

References

Araújo, L.S., Keesman, K.J., Goddek, S., 2021. Making aquaponics a business: a framework. Water (Switzerland) 13 (21), 2978. Available at https://doi.org/10.3390/w13212978.

Baltrusaitis, T., Ahuja, C., Morency, L.P., 2019. Multimodal machine learning: a survey and taxonomy. IEEE Transactions on Pattern Analysis and Machine Intelligence 41 (2), 423–443. Available at https://doi.org/10.1109/TPAMI.2018.2798607.

Chaudhari, A., Bhatt, C., Krishna, A., Travieso-González, C.M., 2023. Facial emotion recognition with inter-modality-attention-transformer-based self-supervised learning. Electronics 12 (2), 288. Available at https://doi.org/10.3390/electronics12020288.

Chen, Y., Li, C., Ghamisi, P., Jia, X., Gu, Y., 2017. Deep fusion of remote sensing data for accurate classification. IEEE Geoscience and Remote Sensing Letters 14 (8), 1253–1257. Available at https://doi.org/10.1109/lgrs.2017.2704625.

David, L.H., Pinho, S.M., Agostinho, F., Costa, J.I., Portella, M.C., Keesman, K.J., et al., 2022. Sustainability of urban aquaponics farms: an emergy point of view. Brazilian Journal of Cleaner Production 331. Available at https://doi.org/10.1016/j.jclepro.2021.129896.

Du, Z., Xu, X., Bai, Z., Liu, X., Hu, Y., Li, W., et al., 2023. Feature fusion strategy and improved GhostNet for accurate recognition of fish feeding behavior. Computers Electronics Agriculture 214, 108310. Available at https://doi.org/10.1016/j.compag.2023.108310.

Feng, S., Yang, X., Liu, Y., Zhao, Z., Liu, J., Yan, Y., et al., 2022. Fish feeding intensity quantification using machine vision and a lightweight 3D ResNet-GloRe network. Aquacultural Engineering 98, 102244. Available at https://doi.org/10.1016/j.aquaeng.2022.102244.

Gümüş, E., Yılayaz, A., Kanyılmaz, M., Gümüş, B., Balaban, M., 2021. Evaluation of body weight and color of cultured European catfish (*Silurus glanis*) and African catfish (*Clarias gariepinus*) using image analysis. Turkish Aquacultural Engineering 93. Available at https://doi.org/10.1016/j.aquaeng.2021.102147.

Höfener, H., Homeyer, A., Förster, M., Drieschner, N., Schildhaus, H.U., Hahn, H.K., 2019. Automated density-based counting of FISH amplification signals for HER2 status assessment. Computer Methods and Programs in Biomedicine 173, 77–85. Available at https://doi.org/10.1016/j.cmpb.2019.03.006.

Hao, Y., Yin, H., Li, D., 2022. A novel method of fish tail fin removal for mass estimation using computer vision. Computers and Electronics in Agriculture 193, 106601. Available at https://doi.org/10.1016/j.compag.2021.106601.

Li, D., Wang, Z., Wu, S., Miao, Z., Du, L., Duan, Y., 2020. Automatic recognition methods of fish feeding behavior in aquaculture: a review. Aquaculture 528, 735508. Available at https://doi.org/10.1016/j.aquaculture.2020.735508.

Liu, H., Liu, T., Gu, Y., Li, P., Zhai, F., Huang, H., et al., 2021. A high-density fish school segmentation framework for biomass statistics in a deep-sea cage. Ecological Informatics 64, 101367. Available at https://doi.org/10.1016/j.ecoinf.2021.101367.

Miličić, V., Thorarinsdottir, R., Santos, M.D., Hančič, M.T., 2017. Commercial aquaponics approaching the European market: to consumers' perceptions of aquaponics products in Europe. Slovenia Water (Switzerland) 9 (2). Available at https://doi.org/10.3390/w9020080.

Palm, H.W., Knaus, U., Kotzen, B., 2024. Aquaponics nomenclature matters: it is about principles and technologies and not as much about coupling. Reviews in Aquaculture 16 (1), 473–490. Available at https://doi.org/10.1111/raq.12847.

Saberioon, M., Gholizadeh, A., Cisar, P., Pautsina, A., Urban, Jan, 2017. Application of machine vision systems in aquaculture with emphasis on fish: state-of-the-art and key issues. Reviews in Aquaculture 9 (4), 369–387. Available at https://doi.org/10.1111/raq.12143.

Shi, C., Wang, Q., He, X., Zhang, X., Li, D., 2020. An automatic method of fish length estimation using underwater stereo system based on LabVIEW. Computers and Electronics in Agriculture 173. Available at https://doi.org/10.1016/j.compag.2020.105419.

Silva, J.F., Hamilton, S., Rocha, J.V., Borie, A., Travassos, P., Soares, R., et al., 2019. Acoustic characterization of feeding activity of *Litopenaeus vannamei* in captivity. Aquaculture 501, 76–81. Available at https://doi.org/10.1016/j.aquaculture.2018.11.013.

Steyaert, S., Pizurica, M., Nagaraj, D., Khandelwal, P., Hernandez-Boussard, T., Gentles, A.J., et al., 2023. Multimodal data fusion for cancer biomarker discovery with deep learning. Nature Machine Intelligence 5 (4), 351–362. Available at https://doi.org/10.1038/s42256-023-00633-5.

Taha, M.F., ElMasry, G., Gouda, M., Zhou, L., Liang, N., Abdalla, A., et al., 2022. Recent advances of smart systems and Internet of Things (IoT) for aquaponics automation: a comprehensive overview. Chemosensors 10 (8), 303. Available at https://doi.org/10.3390/chemosensors10080303.

Wang, Y., Yu, X., Liu, J., An, D., Wei, Y., 2022. Dynamic feeding method for aquaculture fish using multi-task neural network. Aquaculture 551, 737913. Available at https://doi.org/10.1016/j.aquaculture.2022.737913.

Wei, M., Lin, Y., Chen, K., Su, W., Cheng, E., 2020. Study on feeding activity of litopenaeus vannamei based on passive acoustic detection. IEEE Access 8, 156654–156662. Available at https://doi.org/10.1109/access.2020.3019529.

Xu, L., Yu, H., Qin, H., Chai, Y., Yan, N., Li, D., et al., 2024. Digital twin for aquaponics factory: analysis, opportunities, and research challenges. IEEE Transactions on Industrial Informatics 20 (4), 5060–5073. Available at https://doi.org/10.1109/tii.2023.3331135.

Yang, L., Yu, H., Cheng, Y., Mei, S., Duan, Y., Li, D., et al., 2021. A dual attention network based on efficientNet-B2 for short-term fish school feeding behavior analysis in aquaculture. Computers and Electronics in Agriculture 187, 106316. Available at https://doi.org/10.1016/j.compag.2021.106316.

Yanes, A.R., Martinez, P., Ahmad, R., 2020. Towards automated aquaponics: a review on monitoring, IoT, and smart systems. Journal of Cleaner Production 263. Available at https://doi.org/10.1016/j.jclepro.2020.121571.

Zhou, C., Zhang, B., Lin, K., Xu, D., Chen, C., Yang, X., et al., 2017. Near-infrared imaging to quantify thc feeding behavior of fish in aquaculture. Computers and Electronics in Agriculture 135, 233–241. Available at https://doi.org/10.1016/j.compag.2017.02.013.

Zhou, C., Xu, D., Chen, L., Zhang, S., Sun, C., Yang, X., et al., 2019. Evaluation of fish feeding intensity in aquaculture using a convolutional neural network and machine vision. Aquaculture 507, 457–465. Available at https://doi.org/10.1016/j.aquaculture.2019.04.056.

Zeng, Y., Yang, X., Pan, L., Zhu, W., Wang, D., Zhao, Z., et al., 2023. Fish school feeding behavior quantification using acoustic signal and improved swin transformer. Computers and Electronics in Agriculture 204, 107580. Available at https://doi.org/10.1016/j.compag.2022.107580.

Zhang, T., Yang, Y., Liu, Y., Liu, C., Zhao, R., Li, D., et al., 2024. Fully automatic system for fish biomass estimation based on deep neural network. Ecological Informatics 79, 102399. Available at https://doi.org/10.1016/j.ecoinf.2023.102399.

Zhao, S., Zhang, S., Liu, J., Wang, H., Zhu, J., Li, D., et al., 2021. Application of machine learning in intelligent fish aquaculture: a review. Aquaculture 540, 736724. Available at https://doi.org/10.1016/j.aquaculture.2021.736724.

Zhang, L., Li, W., Liu, C., Zhou, X., Duan, Q., 2020. Automatic fish counting method using image density grading and local regression. Computers and Electronics in Agriculture 179, 105844. Available at https://doi.org/10.1016/j.compag.2020.105844.

Zion, B., 2012. The use of computer vision technologies in aquaculture –a review. Computers and Electronics in Agriculture 88, 125–132. Available at https://doi.org/10.1016/j.compag.2012.07.010.

CHAPTER

Equipment and energy consumption control system

7

Daoliang Li[1,2], Bingxiong Wang[1,2] and Quanwu Ge[1,2]

[1]*National Innovation Center for Digital Fishery, China Agricultural University, Beijing, P.R. China,*

[2]*College of Information and Electrical Engineering, China Agricultural University, Beijing, P.R. China*

7.1 Main electrical equipment

Fig. 7.1 presents a schematic diagram of a typical aquaponics system (Zheng et al., 2024). The electrical power sources for the aquaponics system include the main grid and rooftop photovoltaic panels. In terms of electrical equipment, microfilters, UV lamps, and water pumps are indispensable in the wastewater treatment system, ensuring high water quality. Aerators provide the necessary oxygen content, which is crucial for the health of the fish in the tanks. Heat pumps, heating rods, and evaporative cooling pad systems (working in conjunction with wet curtains and ventilators) serve as sources of heating and cooling for indoor air and water temperature control in the fish tanks. Additional electrical equipment in the aquaponics system includes irrigation pumps and baiting traps. The energy management system communicates with each component of the system via a communication unit to facilitate energy management and scheduling.

7.1.1 Illumination lamps

In the agricultural field, lighting fixtures play a crucial role, especially in controlling the plant growth environment. This type of lighting is commonly referred to as illumination lamps as shown in Fig. 7.2, usually employing LED technology capable of providing different wavelengths of light. During seasons with insufficient sunlight, supplementary lighting ensures adequate illumination, aiding in the healthy growth of crops. It is particularly important during plant propagation and seedling stages, as it can accelerate seed germination and seedling growth, improving uniformity and quality of growth.

7.1.2 Internal circulation fan

Internal circulation fans, as depicted in Fig. 7.3, are essential devices for maintaining stable climate conditions and optimizing the plant growth environment.

Smart Aquaponic Farming. DOI: https://doi.org/10.1016/B978-0-443-28945-3.00003-7

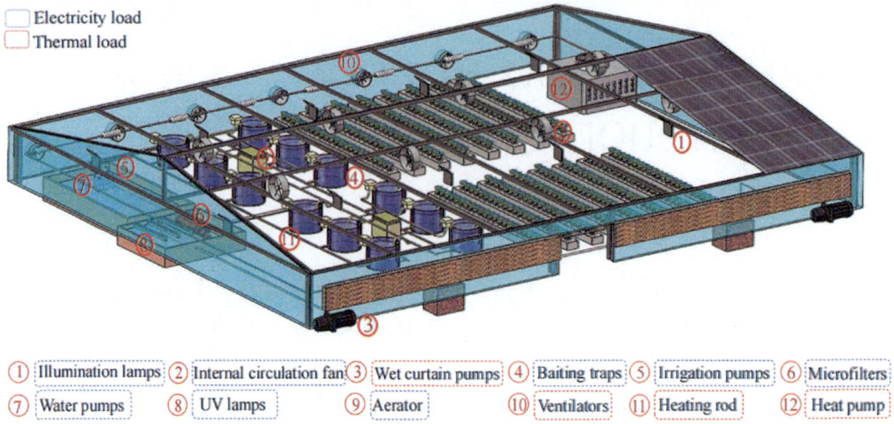

FIGURE 7.1

A typical aquaponics system.

From Zheng, Y., Zhao, W., Varga, M., Li, D., 2024. Stochastic energy management of large industrial-scale aquaponics considering robust optimization-based demand response program. Applied Energy, 374. https://doi.org/10.1016/j.apenergy.2024.123982.

FIGURE 7.2

Illumination lamps.

These fans promote even airflow within the greenhouse, preventing the formation of localized hot or cold spots and ensuring uniform temperature distribution throughout the space. By enhancing air circulation, internal circulation fans help regulate humidity levels within the greenhouse. Good airflow prevents excessive humidity, reducing the risk of diseases and providing a more suitable growth environment for plants.

FIGURE 7.3

Internal circulation fans.

Plants consume carbon dioxide during their growth process. Internal circulation fans facilitate the mixing of gases, helping to maintain an even distribution of carbon dioxide within the greenhouse, which is crucial for promoting photosynthesis and healthy plant growth. Proper air circulation reduces the retention of dew on plant leaves, lowering the chances of fungal and other pathogen growth, thereby protecting crops from diseases.

By simulating natural wind conditions, internal circulation fans also help strengthen the structure of plant stems and leaves, increasing their mechanical strength and improving the plants' ability to adapt to changes in the external environment.

7.1.3 Wet curtain pumps

The wet curtain and pump system, as seen in Fig. 7.4, is a crucial component of greenhouse cooling technology, primarily used to lower the temperature inside the greenhouse through the principle of evaporative cooling. This system includes wet pads (or wet curtains) and pumps, where the pumps continuously supply water to the wet pads, ensuring they remain moist. The system takes advantage of the physical property that evaporation requires heat absorption to cool the air. When the hot air inside the greenhouse passes through the wet pads, the water evaporates, absorbing heat and thereby cooling the air.

FIGURE 7.4

The wet curtain and pump system.

This system is vital for maintaining the temperature inside the greenhouse in high-temperature environments, especially during summer or in warmer regions. By adjusting the water supply and frequency from the pumps, the humidity and evaporation intensity of the wet pads can be controlled, allowing for temperature regulation within the greenhouse. Compared to traditional air conditioning systems, wet pad and pump systems consume less energy and have a smaller environmental impact, making them popular in sustainable agricultural practices.

The wet pad and pump system is relatively simple, easy to install, and maintain. The design of the pumps and wet pads typically considers durability and ease of maintenance for long-term operation. By increasing the humidity of the air inside the greenhouse, the wet pad and pump system also improves the overall air quality needed for plant growth, helping plants grow better and reducing some plant issues caused by dryness.

7.1.4 Baiting traps

In aquaponics systems, baiting traps (as indicated in Fig. 7.5) are devices used to control pests and protect the health of plants and fish within the system. The working principle involves using specific baits, such as food, pheromones, or light sources, to attract pests into the trap, where they are then captured or eliminated.

FIGURE 7.5

Baiting traps.

The structure of baiting traps mainly includes a bait device, a capture container, and an outer casing.

The bait device uses different bait materials, such as sweeteners, pheromones, or light sources, based on the habits of the target pests. The capture container collects and confines the attracted pests, preventing them from escaping. The outer casing protects the internal structure, ensuring the effectiveness of the baiting and capturing process.

The primary functions of baiting traps in aquaponics systems include pest control, reducing pesticide use, and protecting plant health. By attracting and capturing pests, baiting traps effectively reduce plant damage and lower the risk of disease spread, ensuring healthy plant growth. By decreasing pest numbers, baiting traps help reduce reliance on chemical pesticides, maintaining the ecological balance and safety of the system.

In addition, the use of baiting traps can monitor the types and numbers of pests, allowing managers to take appropriate control measures in a timely manner, enhancing the overall efficiency and sustainability of the system. In summary, baiting traps in aquaponics are crucial tools for ensuring the health of plants and fish within the system. Through effective pest control and reduced pesticide use, they promote a positive ecological cycle and sustainable development.

7.1.5 Irrigation pumps

In aquaponics systems, the irrigation pump (as shown in Fig. 7.6) is a crucial device for transporting and circulating water, ensuring efficient water flow between the fish tank and the plant cultivation area. Its working principle involves using an electric

FIGURE 7.6

Irrigation pump.

motor to drive the pump, drawing water from the fish tank, filtering it, enriching it with nutrients, and then delivering it to the plant cultivation area before returning it to the fish tank, thus achieving water recirculation.

The structure of the irrigation pump mainly includes an electric motor, pump body, inlet and outlet pipes, and a filtration device. The electric motor provides the power to rotate the impeller or piston inside the pump, generating water flow. The pump body is designed with corrosion-resistant and wear-resistant materials to ensure long-term reliability. The inlet and outlet pipes connect the fish tank and the plant cultivation area, ensuring smooth water flow. The filtration device removes solid impurities and harmful substances from the water, protecting the health of both the pump and plant roots.

The main functions of the irrigation pump in aquaponics include water circulation, nutrient delivery, and oxygen supply. By continuously circulating the water, the irrigation pump transports water from the fish tank to the plant cultivation area, where fish waste and organic matter in the water serve as nutrient sources for the plants, promoting their growth. The water, after being absorbed and purified by the plants, flows back to the fish tank, providing a clean environment for the fish. The operation of the irrigation pump also increases the dissolved oxygen (DO) content in the water, enhancing the water's self-purification ability and improving the growth conditions for both fish and plants.

Due to the high efficiency and reliability of the irrigation pump, aquaponics systems can achieve water resource recycling and positive ecological interactions, making it an important sustainable development technology in modern agriculture.

7.1.6 Microfilters

A microfilter (as depicted in Fig. 7.7) is a device specifically designed to treat and purify water by removing fine suspended particles and pollutants primarily through physical filtration. The basic principle involves using microfiltration membranes or screens, with pore sizes small enough to block suspended solids, organic particles, and bacteria, while allowing water molecules and certain dissolved substances to pass through.

The core of the device is the microfiltration membrane, typically made from durable polymer materials with pore sizes ranging from 0.1 to 10 micrometers. This effectively retains suspended solids, organic particles, and bacteria while permitting water molecules and certain dissolved substances to pass through. The main

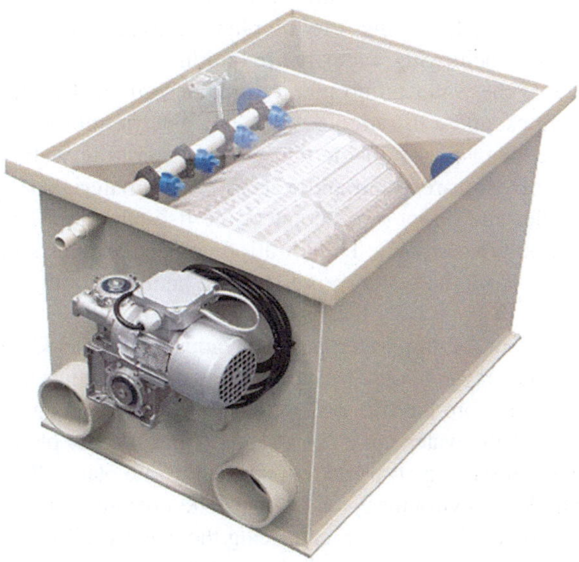

FIGURE 7.7

Microfilter.

components of a microfilter include the microfiltration membrane, a pump to drive water flow, inlet and outlet pipes, pressure gauges, and a control system. In addition, it is equipped with a cleaning system, with common cleaning methods including backwashing and chemical cleaning, to maintain filtration efficiency and extend the membrane's lifespan. This design ensures not only filtration efficiency but also ease of maintenance and operation.

In terms of functionality, microfilters are primarily used to remove particulate matter, control pathogens, support recirculating water systems, and enhance water quality stability. They help reduce the load of suspended matter and microorganisms in the water, lower the risk of disease transmission, and reduce the need for water replacement in recirculating aquaculture systems. By continuously filtering and recirculating the water, they save resources and minimize environmental impact. The use of microfilters improves overall aquaculture efficiency and biological health, making them a crucial component of modern aquaculture technology.

7.1.7 Water pumps

In aquaponics systems, the water pump (as presented in Fig. 7.8) is a crucial device for transporting and circulating water, ensuring efficient water flow between the fish tank and the plant cultivation area. Its working principle involves using an electric motor to drive the pump, drawing water from the fish tank, filtering it, enriching it with nutrients, and then delivering it to the plant cultivation area before returning it to the fish tank, thus achieving water recirculation.

The structure of the water pump mainly includes an electric motor, pump body, inlet and outlet pipes, and a filtration device. The electric motor provides the power to rotate the impeller or piston inside the pump, generating water flow. The pump body is made of corrosion-resistant and wear-resistant materials to ensure long-term reliability. The inlet and outlet pipes connect the fish tank and the plant cultivation area, ensuring smooth water flow. The filtration device removes solid impurities and harmful substances from the water, protecting the health of both the pump and plant roots.

The main functions of the water pump in aquaponics include water circulation, nutrient delivery, and oxygen supply. By continuously circulating the water, the water pump transports water from the fish tank to the plant cultivation area, where fish waste and organic matter in the water serve as nutrient sources for the plants, promoting their growth. The water, after being absorbed and purified by the plants, flows back to the fish tank, providing a clean environment for the fish. The operation of the water pump also increases the DO content in the water, enhancing the water's self-purification ability and improving the growth conditions for both fish and plants.

Due to the high efficiency and reliability of the water pump, aquaponics systems can achieve water resource recycling and positive ecological interactions, making it an important sustainable development technology in modern agriculture.

FIGURE 7.8

Water pumps.

7.1.8 Ultraviolet lamps

Ultraviolet lamps (UV lamps), as shown in Fig. 7.9, used in aquaculture are devices that utilize ultraviolet radiation to purify water and air, widely applied to enhance the sanitary conditions and biological health of the farming environment. The working principle involves ionizing mercury vapor to produce ultraviolet light, especially in the UVC band (200–280 nm), which has strong germicidal capabilities. UVC light can damage the DNA and RNA of microorganisms, effectively killing or inhibiting their reproduction.

The structure of the UV lamp mainly includes a lamp tube made of quartz glass or special UV-transmitting glass, electrodes, mercury vapor, and a power supply unit. The lamp tube is filled with an appropriate amount of mercury, which produces UV light when an electric arc passes through it. The power supply unit provides the voltage and current needed to start and maintain the operation of the lamp tube, usually including a ballast and starter. In addition, protective covers and brackets are used to install and fix the lamp tube. Some models also come with reflectors to increase the efficiency of UV light utilization.

FIGURE 7.9

Ultraviolet lamps.

In aquaculture, the main functions of UV lamps include:

Water sterilization: UVC light kills bacteria, viruses, and other pathogens in the water, ensuring water quality safety.
Air purification: Eliminates microorganisms in the air, improving air quality.
Surface disinfection: Disinfects the surfaces of aquaculture equipment and facilities to prevent the spread of pathogens.
Due to their high efficiency, lack of chemical residues, and ease of operation, UV lamps play a crucial role in aquaculture, significantly improving the sanitary level of the farming environment and the health of farmed organisms.

7.1.9 Aerator

Aerators (as depicted in Fig. 7.10) used in aquaculture are devices designed to increase the DO content in water, ensuring the healthy growth of fish and other aquatic organisms in the farming environment. Their working principle involves mechanically introducing air or oxygen into the water to enhance its oxygen content, thereby improving the oxygen supply in the water.

The structure of an aerator mainly includes an electric motor, impeller or diffuser, air intake pipes, and floats or supports. The electric motor provides the driving power to operate the aeration device. The impeller or diffuser, through rotation or vibration, introduces air into the water, creating bubbles or water movement to promote the mixing of air and water. The air intake pipes bring air from the external environment into the device, typically connecting at the impeller

FIGURE 7.10

Aerator.

or diffuser. The floats or supports are used to fix and support the aerator, ensuring its stable position in the water.

In aquaculture, the main functions of aerators include increasing DO levels, improving water quality, promoting water circulation, and reducing diseases. By introducing air or pure oxygen into the water, aerators significantly increase the concentration of DO, meeting the oxygen needs of fish and other aquatic organisms, and promoting their healthy growth. The aeration process helps decompose organic matter in the water, reducing the accumulation of harmful substances and preventing eutrophication, thereby maintaining a good water quality environment.

The operation of aerators induces water circulation, preventing water stratification, and evenly distributing DO and farming waste, thereby improving the overall quality of the aquaculture water. Adequate oxygen supply can enhance the immune system of aquatic organisms, reduce the incidence of diseases, and increase farming efficiency. Due to their high efficiency, ease of operation, and crucial role in the health of aquatic organisms, aerators have become indispensable equipment in modern aquaculture.

7.1.10 Ventilators

Ventilators (as shown in Fig. 7.11) used in aquaculture are devices designed to improve air quality and regulate temperature in the farming environment, ensuring the healthy growth of aquatic organisms and enhancing production efficiency. Their working principle involves mechanically forcing air movement to increase the frequency of air exchange, thereby removing waste gases, moisture, and excess heat, while introducing fresh air to maintain good ventilation in the farming environment.

The structure of a ventilator mainly includes an electric motor, fan blades, housing, and air inlets and outlets. The electric motor provides power, driving the fan blades to rotate and generate airflow. The fan blades, which can be axial or centrifugal depending on the design, are responsible for drawing in or expelling air. The housing protects the internal components and guides the direction of airflow, while the air inlets and outlets facilitate the intake and exchange of air.

In aquaculture, the main functions of ventilators include improving air quality, regulating temperature and humidity, promoting gas exchange, and reducing disease. By forcing air movement, ventilators effectively remove harmful gases

FIGURE 7.11

Ventilators.

such as carbon dioxide and ammonia from the farming environment, keeping the air fresh and ensuring the normal respiration of aquatic organisms. Ventilators can adjust temperature and humidity, preventing biological stress and diseases caused by excessive temperature or humidity.

Ventilators enhance the exchange of oxygen and carbon dioxide in the air, increasing the DO content in the water, which benefits the growth and metabolism of aquatic organisms. By improving air quality and environmental conditions, ventilators reduce the incidence of diseases, boosting farming efficiency and production. Due to their crucial role in environmental control and biological health maintenance, ventilators have become indispensable equipment in modern aquaculture.

7.1.11 Heating rod

Heating rods (as presented in Fig. 7.12) used in aquaculture are devices designed to regulate the temperature of the water, ensuring that aquatic organisms can grow and reproduce within an optimal temperature range. Their working principle involves converting electrical energy into heat energy through heating elements, which transfer the heat to the water, thereby raising the water temperature.

The structure of a heating rod mainly includes a heating element, a thermostat, a protective casing, and a power connection device. The heating element, usually made of resistive material, efficiently converts electrical energy into heat. The thermostat detects and regulates the water temperature, ensuring that it remains within the set range. The protective casing, made of heat-resistant and waterproof

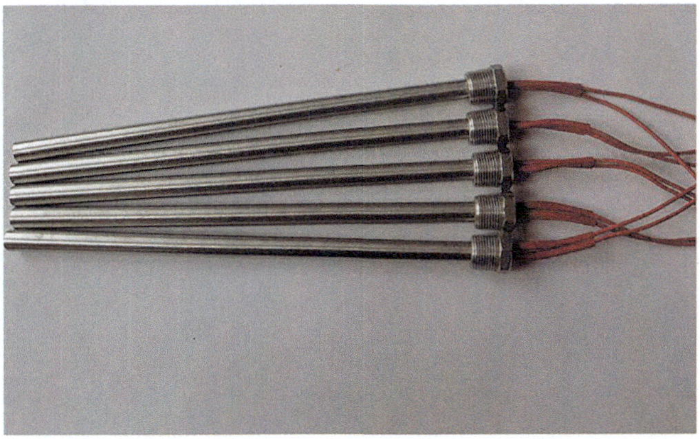

FIGURE 7.12

Heating rods.

materials, ensures the safe operation of the device and prevents electric shocks. The power connection device provides a stable power supply.

In aquaculture, the main functions of heating rods include temperature regulation, maintaining constant temperatures, and promoting growth and development. By raising the water temperature, heating rods ensure that aquatic organisms can grow under suitable temperature conditions, effectively preventing slow growth or disease caused by low water temperatures during cold seasons or in low-temperature environments. The thermostat of the heating rod can monitor the water temperature in real-time and automatically adjust the heating power as needed to maintain a constant water temperature, providing a stable growth environment for aquatic organisms.

Suitable water temperatures help enhance the metabolism and immune system of aquatic organisms, promoting their growth and development and improving farming efficiency. Due to their crucial role in temperature control and biological health maintenance, heating rods have become indispensable equipment in modern aquaculture.

7.1.12 Heat pump

Heat pumps (as depicted in Fig. 7.13) for aquaculture are highly efficient devices used to regulate and maintain air temperatures, ensuring that organisms grow and reproduce within optimal temperature ranges. Their operating principle is based on the reverse Carnot cycle, which involves transferring heat from one area to another to achieve heating or cooling. A typical heat pump system includes a compressor, evaporator, condenser, expansion valve, and circulation pump. The compressor

FIGURE 7.13

Heat pumps.

raises the temperature and pressure of the refrigerant by compressing it. The heated refrigerant then releases heat in the condenser, which is used to warm the water. Subsequently, the refrigerant absorbs heat from the cooler environment in the evaporator, thus cooling down again. The expansion valve regulates the flow and pressure of the refrigerant, ensuring efficient system operation, while the circulation pump drives water through the heat exchanger for heating or cooling. Heat pumps can precisely adjust air temperatures through their heating or cooling functions, based on environmental temperature demands, ensuring that organisms grow in suitable conditions. Compared to traditional heating or cooling equipment, heat pumps have a higher energy efficiency ratio, significantly reducing energy consumption and operational costs. Heat pump systems provide stable water temperatures, avoiding temperature fluctuations that could cause stress reactions in organisms, promoting their growth and enhancing immunity, and reducing the incidence of diseases.

7.2 Fault diagnosis

In aquaponics systems, accurate fault diagnosis is crucial for maintaining stable system operation. Any fault could lead to reduced system efficiency and potentially cause irreversible damage to the ecosystem. Therefore, fault diagnosis requires real-time and online processing capabilities, as well as high sensitivity, to accurately diagnose and locate faults without shutting down or disassembling the system.

The essence of fault diagnosis in aquaponics systems lies in fault classification and localization based on fault pattern recognition. Thus, defining fault patterns is a prerequisite for fault diagnosis. The operating condition data from various nodes form the basis for defining these fault patterns. Due to the involvement of data from multiple types of equipment, multisource data is a characteristic of IoT measurement and control cluster operating conditions. Integrating multisource data to define standardized structures for fault characterization patterns is the main objective of this chapter.

A pattern is a quantitative and structural description of the subject under study, reflecting differences between entities through a set of meaningful features. A pattern class is a collection of patterns with common characteristics. Abnormal values of pattern variables form symptoms, which constitute the fault pattern and serve as the basis for distinguishing between equipment conditions. Symptoms are composed of a collection of indicators, each reflecting the characteristics of a condition through statistical features (such as peak, mean, variance), temporal features, and functional relationships. Thus, the identification of symptom patterns defines the fault patterns and forms the basis for distinguishing different faults.

The fundamental hardware in the system is divided into two categories: sensors and actuators. The sensors include DO sensors and temperature sensors. The actuators comprise system-on-chip core boards, illumination lighting, internal circulation fans, wet curtain pumps, baiting traps, irrigation pumps, microfilters, water pumps, UV lamps, aerators, ventilators, heating rods, and heat pumps. In addition, power supply components such as voltage stabilizers, cables, batteries, solar panels, small wind turbines, small diesel generators, and storage batteries are involved. In this text, these power supply components are categorized under direct current (DC) and alternating current (AC) supplies. The aforementioned sensors are connected to data collectors, and the actuators are connected to distribution control cabinets, forming a measurement and control cluster network. This configuration facilitates the efficient management and operation of the system by ensuring that all components are integrated and can communicate effectively within the network.

The operational indices for each measurement and control cluster include sensor electrode output signal voltage, sensor readings, power supply voltage, single device heartbeat integrity, gateway online status, communication strength, reset frequency, system-on-chip time, actuator unit power, and circulating water leakage signal. The sensor and power supply indices are further divided based on hardware type into DO sensor indices, temperature sensor indices, AC power supply indices, and DC power supply indices. In addition, corresponding statistical measures can be calculated for each index, including instantaneous values, peak values, means, and variances. Instantaneous values correspond directly to the values recorded at specific times. Peak values represent the maximum values within a defined temporal interval (the number of sequential values is discussed in the analysis below). Means reflect the centroid of the index data series within a specific time interval, indicating the average level of the index. Variances measure the dispersion of the index data series within the interval, indicating how much the data points differ from the mean value. These statistical measures provide a comprehensive understanding of the operational conditions and performance of the measurement and control clusters.

Table 7.1, as referenced, displays various symptom indices for the system, where the symbol "√" denotes a symptom index. This symbol represents the type of statistical measure selected for a particular data type, and the number of these symbols indicates the count of symptom indices. This tabulation helps in systematically identifying and quantifying the statistical measures utilized to evaluate and monitor the conditions of the system based on specific symptom indices. Each "√" reflects a distinct measurement being taken into account, facilitating comprehensive diagnostics and system health monitoring.

Among the indices mentioned, those characterized by peak values, means, and variances represent temporal change features, which are one of the critical criteria for fault diagnosis. This means that periodically, after a specific amount of time, when enough time-series data has accumulated, it provides a more substantial basis for conducting fault diagnosis. In addition, if the intervals between data collection are too long, it may hinder timely troubleshooting; conversely, if these intervals are too short, it may lead to false alarms due to insufficient data, thereby causing unnecessary system overhead. To determine a reasonable temporal interval, it is essential to analyze the trend of index data corresponding to typical faults, as illustrated in Fig. 7.14. This analysis will help identify optimal data collection

Table 7.1 Symptom indicators.

Typology	Type of statistic			
	Prompt value	Peak value	Average value	Variance
Electrode output signal voltage	√			
Sensor acquisition value	√	√		
AC		√	√	
DC	√			√
Single device heartbeat integrity	√			
Gateway online status	√			√
Communications strength	√	√		
Reset frequency			√	
System-on-chip time	√			
Actuator power	√	√		√
Circulating water leakage signal	√			

intervals that balance the need for sufficient diagnostic data against the costs of data processing and the risk of false alarms, thereby enhancing the efficiency and accuracy of the fault diagnosis process.

7.3 Energy control

7.3.1 Equipment characterization in industrial-scale aquaponics

The basis of smart aquaponics farming requires a large number of electrical devices in industrial-scale aquaponics. These equipments are used to ensure precise control of the culture and growing environment and can be categorized into the following three groups based on their operational characteristics: non-stop, periodic, and empirical (as depicted in Fig. 7.15).

Non-stop electrical equipment is equipment whose operation is necessary to maintain the basic work of an aquaponics system, for example, circulating water pump that continuously circulates water through heater, pipe, aquaculture system, and hydroponics system. Periodic electrical equipment refers to equipment that regularly works at a specific time each day, such as bait casting machines and factory lighting systems.

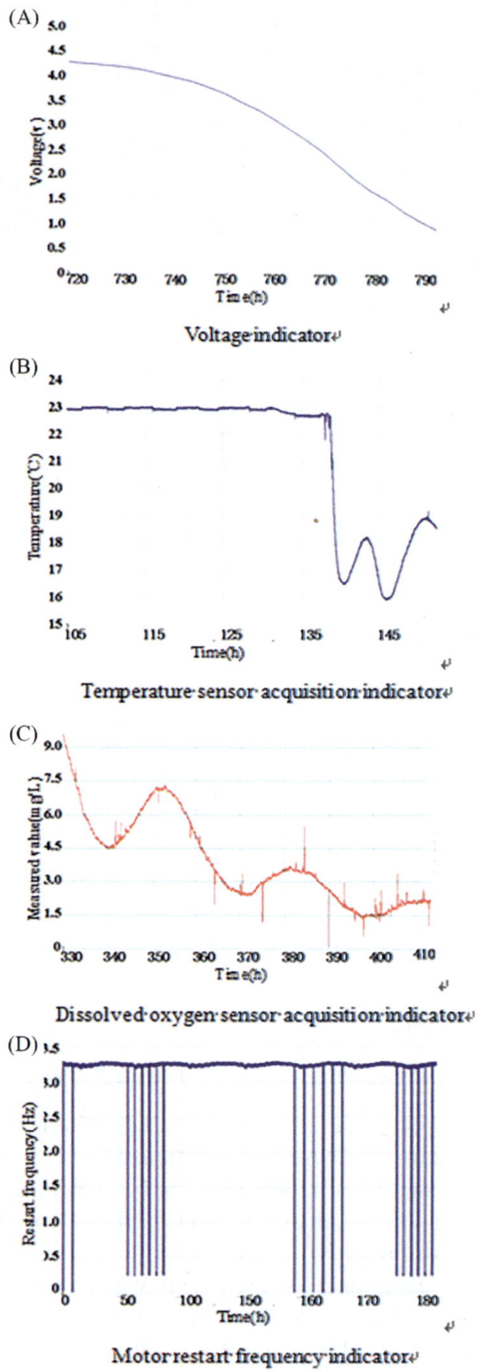

FIGURE 7.14

Indicator data curves corresponding to some of the faults. (A) Voltage indicator; (B)Temperature sensor acquisition indicator; (C) Dissolved oxygen sensor acquisition indicator; (d) Motorrestart frequencyindicator.

Unit	Operational characteristics			Flexibility characteristics			Quantity/ Rated power (kW)
	Work non-stop	Periodic	Empirical	Non - schedulable	Schedulable		
					Time window-shiftable	Triggered "on/off"	
UV lamps	🟢◆	🔴◆	🔴◆	🟢◆	🔴◆	🔴◆	9/0.055
Water pumps	🟢◆	🔴◆	🔴◆	🟢◆	🔴◆	🔴◆	2/3
Microfilter	🔴◆	🟢◆	🔴◆	🟢◆	🔴◆	🔴◆	1/3
Baiting traps	🔴◆	🔴◆	🟢◆	🔴◆	🟢◆	🔴◆	12/0.4
Irrigation pump	🔴◆	🔴◆	🟢◆	🔴◆	🟢◆	🔴◆	1/2.2
Ventilators	🔴★	🔴★	🟢★	🔴★	🔴★	🟢★	10/1.1
Wet curtain pumps	🔴★	🔴★	🟢★	🔴★	🔴★	🟢★	2/0.75
Illumination lamps	🔴◆	🔴◆	🟢◆	🟢◆	🔴◆	🔴◆	14/0.1
Circulation fan	🔴◆	🔴◆	🟢◆	🟢◆	🔴◆	🔴◆	10/0.25
Aerator	🟢◆	🔴◆	🔴◆	🔴◆	🔴◆	🟢◆	1/1.4
Heat pump	🔴★	🔴★	🟢★	🔴★	🔴★	🟢★	3/10
Heating rods	🔴★	🔴★	🟢★	🔴★	🔴★	🟢★	12/2

🟢 applicable 🔴 non applicable ◆ electricity load ★ heat load

FIGURE 7.15

Characterization of electrical equipment in industrial-scale aquaponics.

Zheng, Y., Zhao, W., Varga, M., Li, D., 2024. Stochastic energy management of large industrial-scale aquaponics considering robust optimization-based demand response program. Applied Energy, 374. https://doi.org/10.1016/j.apenergy.2024.123982.

Empirical operation refers to the operation mode formed by the equipment relying on the experience of experts and the actual needs of farming. The electrical equipment based on empirical operation mode includes temperature, DO and other control equipment, for example, the water temperature and DO require a specific suitable range, and the temperature and DO control equipments can be set to switch on and off when they do not meet the requirements.

However, the empirical operation mode can usually be improved by intelligent control strategy design. Intelligent control strategies can be realized to achieve higher accuracy and lower energy consumption under the basic conditions of meeting the needs of fish culture and vegetable cultivation. Therefore, intelligent control strategy design is necessary for intelligent aquaponics.

7.3.2 Energy control system

7.3.2.1 *Energy consumption model*

In aquaponics, DO concentration and water temperature are important aquaculture environmental parameters. To maintain the DO content of the aquaculture tanks, aerators are often used, and for water temperature, cleaner, and more energy-efficient heat pump systems are often used. Mechanism models are commonly used as effective tools for designing control systems. In this subsection, the modeling of DO and water temperature as the main research object is introduced. Next, the common controller designs used in control systems as well as the energy optimization model when combining different energy systems in a smart aquaponics farm.

7.3.2.1.1 Dissolved oxygen

Taking into account the intricacies of an aquaponics, a comprehensive assessment was made of the principal factors influencing the dynamic fluctuations in DO within the culture tank (as illustrated in Fig. 7.16). This encompassed the recirculating water flow (R_{wf}), mechanical aeration (R_{aer}), Surface reaeration (R_{re}), and fish respiration (R_{fish}), where, C is the DO concentration. Subsequently, a differential equation model was formulated to delineate the dynamic variations in DO concentration within the water Eq. (7.1), as expressed below (Chai et al., 2024):

$$\frac{dC}{dt} = R_{wf} + R_{aer} + R_{re} - R_{fish} \tag{7.1}$$

In ordinary circumstances, DO concentration in an aquaponics is predominantly influenced by the aforementioned four factors. Consequently, Eq. (7.1) could serve as a differentially equation model capturing the dynamic variations of DO in the majority of an aquaponics.

7.3.2.1.2 Water temperature

The mathematical model depicts the dynamics of water temperature within aquaponics, comprising five heat exchange processes (Li et al., 2009), and schematic diagram shown in Fig. 7.17. They are solar radiation absorbed by water (Q_{sw}), the latent heat loss from the evaporation of the water (Q_{lw}), the sensible heat transfer between the water surface and the inside air (Q_{vw}), the heat exchange between the system wall and surrounding environment (Q_{wall}), the net thermal radiation between the water surface and the sky (Q_{sky}), and the external

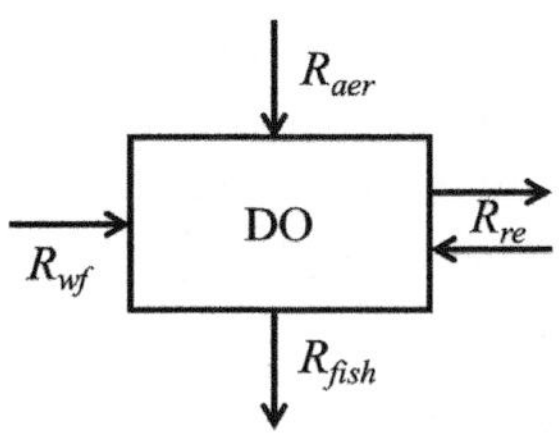

FIGURE 7.16

Schematic diagram of the dynamic balance of DO. *DO*, dissolved oxygen.

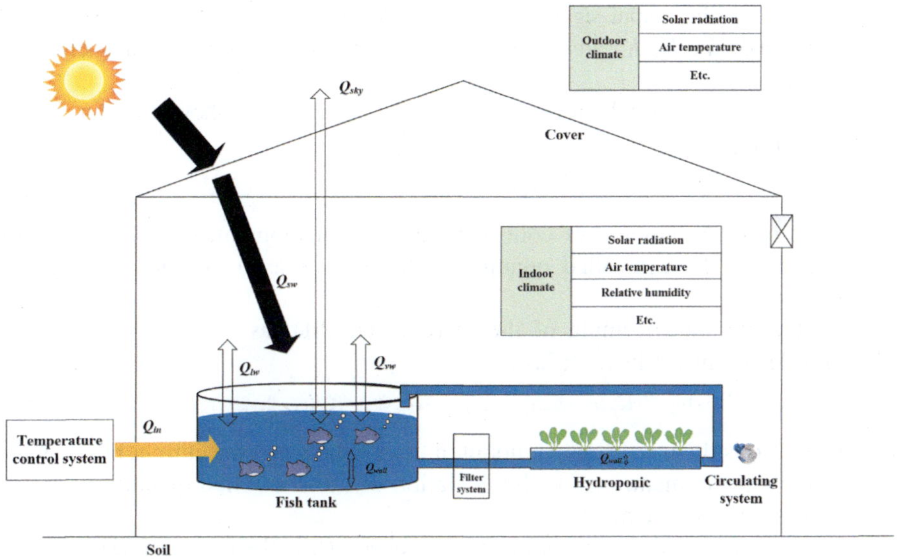

FIGURE 7.17

Schematic diagram of the dynamic balance of water temperature.

temperature control device, such as a heat pump, provides energy to the body of water (Q_{in}), can be defined as Eq. (7.2):

$$c_{p,w}\rho_w V_w \frac{dT_w}{dt} = Q_{sw} - Q_{vw} - Q_{lw} - Q_{wall} + Q_{in} \tag{7.2}$$

The energy balance function for water inside aquaponics is given as:where $c_{p,w}$ is the specific heat capacity of the water (J/(kg°C)), ρ_w is the density of the water (kg·m^{-3}), V_w is the volume of the water (m^3), T_w is the water temperature (°C).

7.3.2.2 Energy control strategy

7.3.2.2.1 Controller design

1. Conventional controller Conventional control strategies include switching control, timing control, proportional-integral-derivative (PID) control, fuzzy control, and so on. Among them, PID controller is a commonly used controller, which has the advantages of simple structure, easy operation, and strong robustness. PID controller is a feedback based controller, and the time domain expression of the PID controller can be expressed as Eq. (7.3):

$$u(t)=K_p\left[e(t) + \frac{1}{K_i}\int_0^t e(t)dt + K_d\frac{de(t)}{dt}\right] \tag{7.3}$$

where, $u(t)$ is the output signal of the controller, $e(t)$ is the current deviation of the system, and Kp, K_i, K_d are the proportional, integral, and differential coefficients, respectively.

The discrete state formula of the positional PID controller (as showed in Fig. 7.18) is shown in Eq. (7.4):

$$u_k = K_p e_k + K_i \sum e + K_d (e_k - e_{k-1}) \tag{7.4}$$

where, u_k is the amount of control at the current moment, e_k and e_{k-1} are the deviation of the controlled quantity to the set value at current and previous moments.

The discrete state formula of the incremental PID (as depicted in Fig. 7.19) controller is shown in Eq. (7.5).

$$\Delta u_k = K_p (e_k - e_{k-1}) + K_i (e_k - e_{k-1}) + K_d (e_k - 2e_{k-1} + e_{k-2}) \tag{7.5}$$

where, Δu_k is the incremental amount of control between the current moment and the previous moment, e_{k-2} is the deviation of the controlled quantity to the set value at first two moments.

where, $u(t)$ is the output signal of the controller, $e(t)$ is the current deviation of the system, and Kp, K_i, K_d are the proportional, integral, and differential coefficients, respectively.

The traditional PID control parameters are difficult to determine and cannot be adjusted online, so there are still some deficiencies in the control accuracy. Therefore, the method of adjusting PID parameters based on fuzzy rules is proposed, and this kind of controller mainly formulates a series of fuzzy rules to adjust PID parameters online according to the experience of experts, so as to improve the adaptive ability of the controller.

In addition, many researchers have combined neural networks with PID controllers, using the powerful self-learning ability of neural networks to adjust PID parameters to improve the adaptive ability of PID, such as back propagation-PID and radial basis

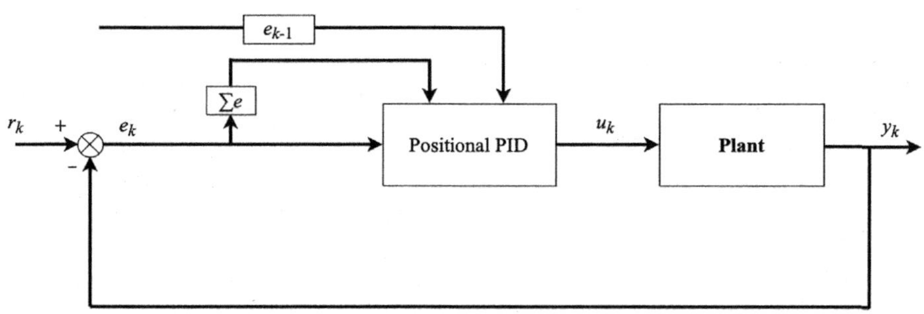

FIGURE 7.18

Schematic diagram of positional PID. *PID*, proportional-integral-derivative.

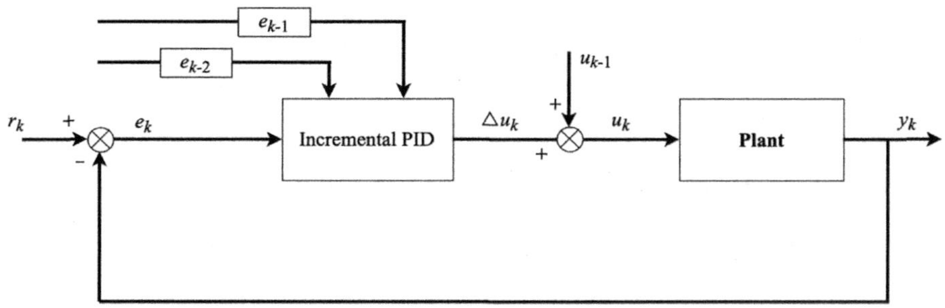

FIGURE 7.19

Schematic diagram of incremental PID. *PID*, proportional-integral-derivative.

function-PID controllers, and fuzzy rule-optimized single-neuron adaptive PID controllers (FL-SN-PID) for precise control (Zhou et al., 2022).

2. Modern controller modern control theory is based on state variables and state space equations. A state space formulation is a series of differential equations that describe the inputs and outputs of a system as well as the state variables, as mentioned in subsection 3.2.1 energy consumption model.
The state equation are shown in Eq. (7.6):

$$\begin{aligned} \dot{x} &= Ax + Bu \\ \dot{y} &= Cx + Bu \end{aligned} \tag{7.6}$$

This system of equations describes the dynamic behavior of the system, where x denotes the state variable, y denotes the system output and u denotes the system input.

In model predictive control (MPC), a model describing the dynamic behavior of the object is needed, and the role of this model is to predict the future dynamics of the system. Predictive models usually include: convolutional models: step response models and impulse response models; 2. mechanistic models: state space models and transfer function models; 3. fuzzy models; 4. artificial neural network (NN) models; 5. chaos (Chaos) models (Chen, 2013). Due to the fact that the differential equations describing the system can be easily rewritten in state-space form, state-space based model-model predictive control is a commonly used approach.

The steps of MPC (as illustrated in Fig. 7.20) are usually divided into: 1. modeling the system; 2. predicting the state of the system at a future time; 3. optimizing the optimal control inputs to the system at a future time.

The predicted state matrix, target matrix, and control matrix for the later p prediction cycles are:

$$\begin{aligned} X_k &= [x_{k,k+1} x_{k,k+2} \cdots x_{k,k+p}]^T \\ R_k &= [r_{k,k+1} r_{k,k+2} \cdots r_{k,k+p}]^T \\ U_k &= [u_k u_{k,k+1} \cdots u_{k,k+p-1}]^T \end{aligned} \tag{7.7}$$

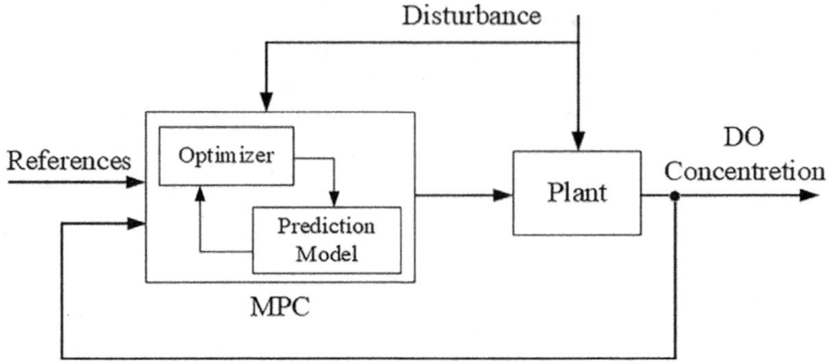

FIGURE 7.20

Structure of model predictive controller.

Chai, F., Ge, Q., Yin, Y., Li, D., Wang, Y., 2024 Modeling and control of DO in recirculating aquaculture systems: a circadian rhythm analysis approach and GSMPC controller. Computers and Electronics in Agriculture, 227. https://doi.org/10.1016/j.compag.2024.109515.

The system's predicted state can be expressed in matrix form as:

$$X_k = \psi x_k - \Theta U_k \tag{7.8}$$

where $\boldsymbol{\Psi}$ and $\boldsymbol{\Theta}$ are the state, input matrix coefficients.

During the optimization step, a cost function needs to be set, such as the common task of tracking a reference target, which requires the input control input error to be minimized, and the cost function is defined as the cumulative error between the predicted state vector and the reference value, subject to the constraints on the control variables, can be described as:

$$J = (X_k - R_k)^T W (X_k - R_k) + U_k^T Y U_k \tag{7.9}$$

where $\boldsymbol{W}$ and $\boldsymbol{Y}$ are the error weight matrix and the control weight matrix, respectively.

Define $\boldsymbol{E}=\boldsymbol{\Psi} x_k - \boldsymbol{R}_k$, and cost function can be shown as:

$$J = U_k^T(\Theta^T W \Theta + Y)U_k + (2E^T W \Theta + Y)U_k + E^T W E \tag{7.10}$$

Define: $\boldsymbol{H}=2(\boldsymbol{\Theta}^T \boldsymbol{W\Theta} + \boldsymbol{Y})$, $f^T=2\boldsymbol{E}^T\boldsymbol{W\Theta}$, the above cost function is a typical quadratic programming problem, and this linearly constrained quadratic optimization problem can be expressed by the standard quadratic programming (QP) problem formulation:

$$\begin{gathered} \min U \tfrac{1}{2} U_k^T H U_k + f^T U_k \\ \text{s.t.} \\ U_{min} \le U_k \le U_{max} \end{gathered} \tag{7.11}$$

In water temperature controller design, researchers noted that traditional threshold range control strategies exhibit imprecision when sensor resolution limitations are present. They proposed resolving this through the trend strategy, which is suitable for both model-free and model-based controllers, such as PID and MPC controllers (Ge et al., 2025). In addition, research has proved that fish respiration has the phenomenon of circadian rhythm, that is, the periodic consumption of DO during the growth of fish. Incorporating such phenomena, the researcher designed the gain-scheduling model predictive control (GSMPC) controller with the main aim to minimize energy consumption by switching the appropriate MPC controllers at the right time while ensuring optimal fish growth (Chai et al., 2024). In this controller, four MPC controllers with different performances were designed (as illustrated in Table 7.2). The rule of thumb for the MPC controller design is: control range = 10%~30% × prediction range. The GSMPC controller switches between the four controllers using gain scheduling signals according to the requirements such as the oxygen supply and the circadian rhythm of the fish.

7.3.2.2.2 Energy optimization model

In areas where natural resources are abundant, local natural resources can be used to reduce farming costs. For example, the heat energy for the trout farms (cold-water fish) using the geothermal energy in Meshginshahr City, located in Sabalan foothills of Iran (GharibiAsl and Abbaspour-Gilandeh, 2019). However, in smart aquaponics farms, in order to ensure a standard production environment to improve the farming efficiency and reduce the pollution emission from farming, renewable energy systems, such as wind, photovoltaic, and heat pumps, are often chosen to be used as the energy supply system for production (Le et al., 2020), Fig. 7.21. In addition, due to the uncertainty of renewable energy sources such as wind and photovoltaic, and the need to minimize operating costs while maintaining production requirements, it is necessary to consider the different energy systems in an integrated manner and to

Table 7.2 Parameter settings of model predictive control (MPC) controllers.

Controller	Sample time (min)	Prediction horizon	Control horizon	Error weight	Control weight	ECR weight
MPC1	10	10	3	100	10	1000
MPC2	10	10	3	100	50	1000
MPC3	10	10	3	100	100	1000
MPC4	10	10	3	10	100	1000

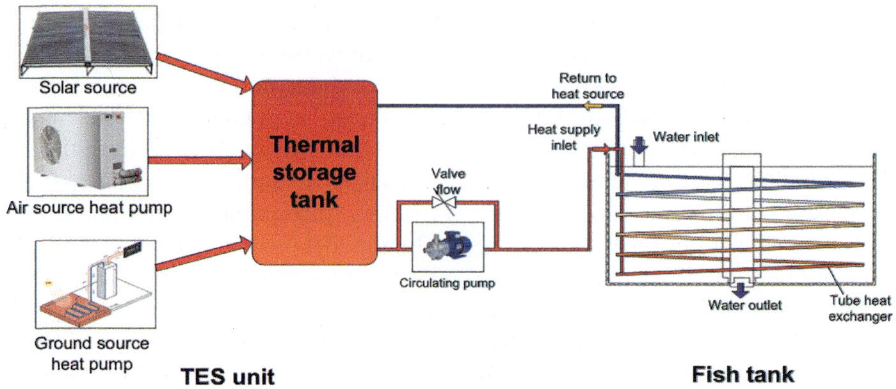

FIGURE 7.21

Different energy systems for temperature control in aquaculture tanks.

Le, A.T., Wang, Y., Wang, L., Ta, V.C., Li, D., 2020. Numerical investigation on a low energy-consumption heating method for recirculating aquaponic systems. Computers and Electronics in Agriculture 169, 105210. https://doi.org/10.1016/j.compag.2019.105210.

perform a rational and optimal scheduling. The minimum cost optimization model for a multi-energy supply system is shown below (Zheng et al., 2024):

$$\begin{gathered} \min J = P_{grid,e} f_e + P_{o,e} f_o \\ \text{s.t.} \\ P_{t,e} = P_{grid,e} + P_{o,e} + P_{rd,e} + P_{s,e} \\ 0 \le P_{s,e} \le P_{s,\max} \end{gathered} \tag{7.12}$$

References

Chai, F., Ge, Q., Yin, Y., Li, D., Wang, Y., 2024. Modeling and control of dissolved oxygen in recirculating aquaculture systems: a circadian rhythm analysis approach and GSMPC controller. Computers and Electronics in Agriculture 227, 109515. Available at https://doi.org/10.1016/j.compag.2024.109515.

Chen, H., 2013. Model predictive control. Science Press.

Ge, Q., Yang, J., Chai, F., Zhu, F., Yang, W., Zhang, Z., et al., 2025. A novel trend threshold range control strategy for enhancing energy efficiency and water temperature control performance in industrial-scale greenhouse aquaculture. Energy 338, 138879. Available at https://doi.org/10.1016/j.energy.2025.138879.

GharibiAsl, S., Abbaspour-Gilandeh, Y., 2019. Evaluation of the direct use of geothermal energy on heat factors required for cold-water fish pisciculture. Aquaculture (Amsterdam, Netherlands) 512, 734291. Available at https://doi.org/10.1016/j.aquaculture.2019.734291.

Le, A.T., Wang, Y., Wang, L., Ta, V.C., Li, D., 2020. Numerical investigation on a low energy-consumption heating method for recirculating aquaponic systems. Computers and Electronics in Agriculture 169, 105210. Available at https://doi.org/10.1016/j.compag.2019.105210.

Li, S., Willits, D.H., Browdy, C.L., Timmons, M.B., Losordo, T.M., 2009. Thermal modeling of greenhouse aquaculture raceway systems. Aquacultural Engineering 41 (1), 1–13. Available at https://doi.org/10.1016/j.aquaeng.2009.04.002.

Zheng, Y., Zhao, W., Varga, M., Li, D., 2024. Stochastic energy management of large industrial-scale aquaponics considering robust optimization-based demand response program. Applied Energy 374. Available at https://doi.org/10.1016/j.apenergy.2024.123982.

Zhou, X., Wang, J., Huang, L., Li, D., Duan, Q., 2022. Modelling and controlling dissolved oxygen in recirculating aquaculture systems based on mechanism analysis and an adaptive PID controller. Computers and Electronics in Agriculture 192, 106583. Available at https://doi.org/10.1016/j.compag.2021.106583.

CHAPTER

Microbial regulatory system

8

Daoliang Li[1,2], Zhuangzhuang Bai[1,2] and Jie Han[1,2]

[1]*National Innovation Center for Digital Fishery, China Agricultural University, Beijing, P.R. China,*

[2]*College of Information and Electrical Engineering, China Agricultural University, Beijing, P.R. China*

8.1 Introduction

Microorganisms in nature are incredibly small organisms, including bacteria, fungi, viruses, and more. As essential components of biodiversity, they are widely distributed across various environments, such as soil, water, air, and even within living organisms. These microorganisms play crucial ecological and biological roles. One of their primary functions is breaking down complex organic matter into simpler inorganic compounds, thereby facilitating the cycling of essential elements like nitrogen and phosphorus. This process ensures that these nutrients can be efficiently absorbed and utilized by plants.

In aquaponics systems, one of the most fundamental processes involves nitrifying bacteria converting ammonia nitrogen into nitrites and then into nitrates (Fig. 8.1). This transformation provides plants with readily absorbable nitrogen nutrients, which are essential for their growth. Moreover, this microbial activity plays a key role in maintaining the stability and balance of the aquaponics ecosystem, ensuring its sustainability and efficiency.

Therefore, microorganisms play an indispensable role in aquaponics systems, influencing water quality, the stability of the aquaculture environment, and the health of aquatic organisms. As aquaculture scales expand, the waste excreted by fish increasingly affects water quality, leading to potential ecological imbalances. One of the most significant consequences is water eutrophication, where excessive nutrients promote the rapid growth of microorganisms, creating an environment rich in nutrients and ideal for microbial proliferation. Microorganisms in aquaponics systems encompass a diverse range, including bacteria, fungi, and viruses, among others. These microorganisms form a complex network of interactions with water bodies, substrates, and cultured organisms, directly or indirectly impacting the stability and productivity of the system. Given their profound influence, a comprehensive understanding of the types and functions of microorganisms in

Smart Aquaponic Farming. DOI: https://doi.org/10.1016/B978-0-443-28945-3.00009-8

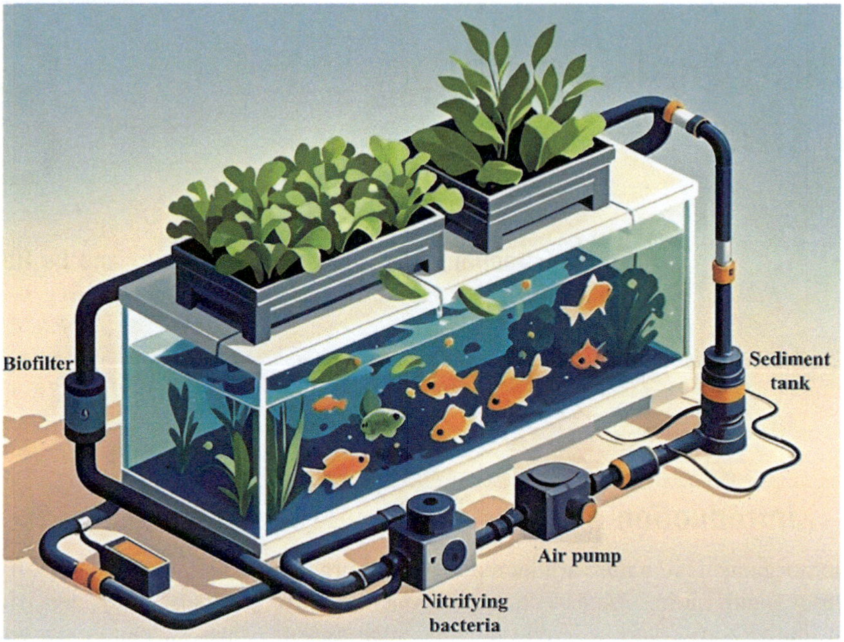

FIGURE 8.1 The aquaponics system.

illustrates the principle of an aquaponics system. The system integrates soilless plant cultivation in the upper section with fish farming in the lower tank, creating a sustainable ecosystem through water recirculation.

aquaculture is essential. Such knowledge can enhance aquaculture efficiency, maintain environmental stability, and promote the health of cultured organisms, ultimately contributing to the sustainability of the aquaponics ecosystem.

8.2 Bacteria

Bacteria play a vital and multifaceted role in aquaponics, encompassing both beneficial and potentially harmful species. Beneficial bacteria contribute significantly to system stability by breaking down organic matter, converting it into essential nutrients that can be readily absorbed by plants. This process not only promotes plant growth but also enhances overall system productivity. Furthermore, beneficial bacteria play a crucial role in key nutrient cycles, such as the nitrogen and phosphorus cycles, facilitating biological nitrogen fixation and organic matter decomposition to help maintain ecological balance and system stability (Kasozi et al., 2021). However, harmful bacteria pose potential threats to aquaponics systems, as they can cause diseases that negatively impact the health of both fish and plants. In addition, the overproliferation of harmful bacteria can lead to ecosystem instability and deteriorate water quality. Therefore, effective management and monitoring of bacterial communities are essential in aquaponics. By optimizing the structure and function of bacterial populations, it is possible to maximize the benefits of beneficial bacteria while minimizing potential risks, ensuring the healthy operation and sustainable development of the system.

Aquaponics systems host a diverse array of bacterial species, each exhibiting distinct forms, growth characteristics, and ecological functions. The following section will introduce some of the most common and essential bacteria, highlighting their roles and significance within the system.

8.2.1 Nitrifying bacteria

Nitrifying bacteria are *Gram-negative*, aerobic microorganisms that play a crucial role in aquaponics systems. They are primarily classified into two groups: nitrite-oxidizing bacteria and nitrate-oxidizing bacteria. These bacteria commonly exhibit rod, spherical, or spiral shapes. Known for their adaptability, they can function at relatively low temperatures, although their optimal growth temperature typically ranges between 15°C and 35°C. The activity and growth of nitrifying bacteria are significantly influenced by pH levels. They thrive best in neutral to alkaline environments, with an optimal pH range of 7.5–8.5. As illustrated in Fig. 8.2, nitrifying bacteria serve as a vital link between fish waste and plant nutrients in aquaponics systems. Through the process of nitrification, ammonia nitrogen (NH_4^+) in circulating water is first oxidized into nitrite (NO_2^-) by nitrite-oxidizing bacteria. Subsequently, nitrite is further oxidized into nitrate (NO_3^-) by nitrate-oxidizing bacteria, making it available for plant absorption and utilization (Zou et al., 2016). This process is essential for maintaining water quality and ensuring the sustainability of the aquaponics ecosystem.

8.2.2 Aeromonas hydrophila

Aeromonas hydrophila is a *Gram-negative*, facultative anaerobic bacterium characterized by a single flagellum, motility, and the absence of both capsules and spores. It thrives within a broad temperature range of 14.0°C–40.5°C, with an optimal growth temperature of 28°C–30°C. This bacterium can survive in pH conditions ranging from 6 to 11, with an optimal pH of 7.2~7.4. In addition, it can tolerate salinity levels between 0% and 4%, with an optimal salinity of 0.5%. *A. hydrophila* is a major pathogenic bacterium in freshwater aquaculture, responsible for severe diseases in various fish species. It is the primary cause of hemorrhagic septicemia in silver carp, red fin disease in eels, vertical scale disease in carp and goldfish, printing disease in silver and bighead carp, and bacterial enteritis in black carp and grass carp. Moreover, it can also lead to septicemia in fish such as salmon

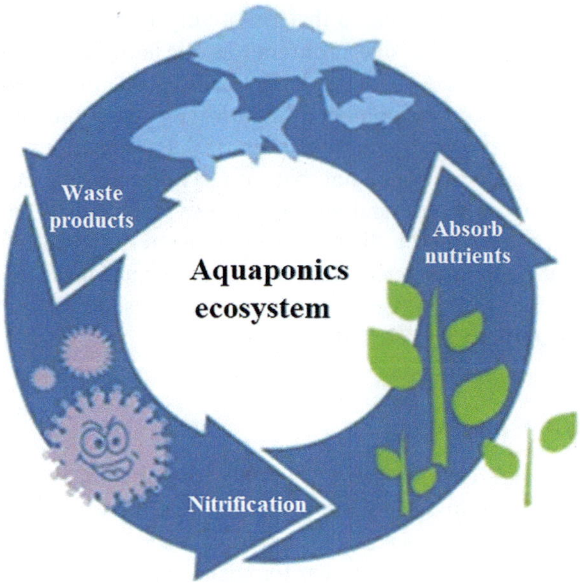

FIGURE 8.2 The role of nitrifying bacteria.

illustrates the critical role of nitrifying bacteria in an aquaponics system. It depicts the biofiltration process where bacteria convert toxic ammonia from fish waste into nitrites and then into nitrates, which serve as a vital nutrient source for the plants.

and trout. Infection typically occurs through the intestinal tract but can also penetrate the body via damaged skin or gills, especially when fish suffer injuries or parasitic infections. Several environmental factors, including water temperature, organic matter content, and stocking density, influence the prevalence and severity of infections. Sudden environmental changes and deteriorating water quality can exacerbate infections, often in combination with other opportunistic pathogens such as *Aeromonas sobria* and Vibrio species, leading to more severe disease outbreaks. Infections caused by *A. hydrophila* are typically acute and highly contagious, often resulting in high mortality rates (Dong et al., 2017). Fig. 8.3 shows *A. hydrophila*-infected fish exhibiting: dermal/subcutaneous congestion and hemorrhage, exophthalmia, abdominal distension with straw-colored or hemorrhagic ascites, hepatosplenomegaly and renomegaly, mottled liver, dark-purple to black spleen, and severe anemia.

8.2.3 Edwardsiella tarda

Edwardsiella tarda belongs to *Gram-negative bacteria*, is a class of facultative anaerobic bacteria, single or in pairs, periflagellate, motile, capsule free, spore free.

FIGURE 8.3 The fish infected with *Aeromonas hydrophila* (Dongtai Fishery Technology Guidance Station, Jiangsu, China).

(A) External appearance showing subcutaneous hemorrhage and hyperemia on the body surface, fins, and gill region; (B) Dissected internal cavity showing tissue hemorrhage, organ discoloration, and pathological damage to visceral organs. Shows a fish infected with *Aeromonas hydrophila*, exhibiting typical clinical signs including subcutaneous hemorrhage, exophthalmia, abdominal swelling with ascites, hepatosplenomegaly, and severe anemia. This gram-negative bacterium is a major pathogen in freshwater aquaculture, causing acute and highly contagious disease outbreaks, often under poor water quality or stressful environmental conditions.

The optimal temperature range for its growth is 25°C~32°C, while it ceases to grow at temperatures above 42°C. This bacterium can thrive in pH conditions ranging from 5.5 to 9 and tolerate salinity levels between 0 and 4%. *E. tarda* is the primary causative agent of edwardsiellosis in fish, a disease that affects a wide range of aquatic species, including goldfish, rainbow trout, largescale salmon, eel, flounder, tilapia, and sea bass. It is one of the more prevalent pathogenic bacteria in aquaculture. While edwardsiellosis can occur throughout the year, it predominantly emerges when water temperatures exceed 15°C, with peak outbreaks occurring between 25°C and 30°C. Fish typically become infected through the ingestion of food contaminated with *E. tarda* or by direct contact with infected fish. The bacterium primarily enters the host through the digestive tract, gills, or damaged skin (Armwood et al., 2022). As depicted in Fig. 8.4, *E. tarda*–infected eels exhibit diffuse cutaneous and fin hemorrhage, a protruding erythematous anus, visceral organomegaly, and congestion. Due to its wide host range and ability to thrive under varying environmental conditions, *E. tarda* poses a significant threat to aquaculture.

8.2.4 Streptococcus agalactis

Streptococcus agalactis, a Gram-positive bacterium, is a facultative anaerobic bacterium with round or oval cells arranged in a chain pattern. It is non-motile, lacks flagella, and does not form spores. This bacterium thrives in salinity levels ranging from 0% to 4%, with an optimal growth temperature between 10°C and 40°C and a preferred pH range of 4.4–7.6. *S. agalactiae* is a significant pathogen in aquaculture, infecting a variety of fish species, including mullet, golden-head bream, sweetfish, eel, and rainbow trout. However, it poses the greatest threat to tilapia, where infections can cause severe damage. The incidence of infection decreases when water temperatures fall below 20°C, but during warmer seasons, outbreaks peak, with mortality lasting for 2–3 weeks. Infected fish typically exhibit systemic congestion and hemorrhaging, often accompanied by intense inflammatory responses. Affected fish show signs of reduced appetite, erratic swimming, body curvature, and lethargy (Haenen et al., 2023). Characterized by a chronic course, the disease progression in infected fish manifests as hepatic hemorrhage, intraluminal

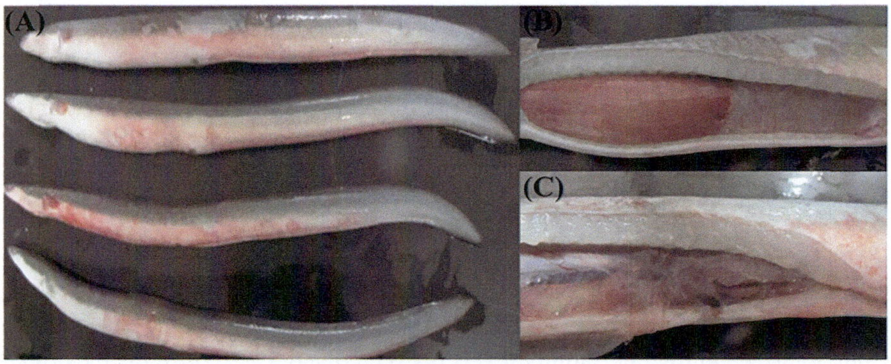

FIGURE 8.4 The fish infected with *Edwardsiella tarda* (Liyang Aquatic Technology Co., Ltd, Guangzhou, China).

(A) External lesions (cutaneous hemorrhage); (B,C) Internal pathological changes (muscle discoloration and visceral congestion). Shows *Edwardsiella tarda*-infected fish exhibiting typical clinical signs including diffuse cutaneous and fin hemorrhage, protruding erythematous anus, and visceral organ congestion.

intestinal purulent exudate, abdominal effusion, exophthalmia, and hemorrhagic lesions at fin bases (Fig. 8.5) (Fan et al., 2025).

8.2.5 Pseudomonas fluorescens

Pseudomonas fluorescens is a *Gram-negative*, aerobic bacterium with a rod-shaped structure and blunt, rounded ends. It thrives at an optimal temperature range of

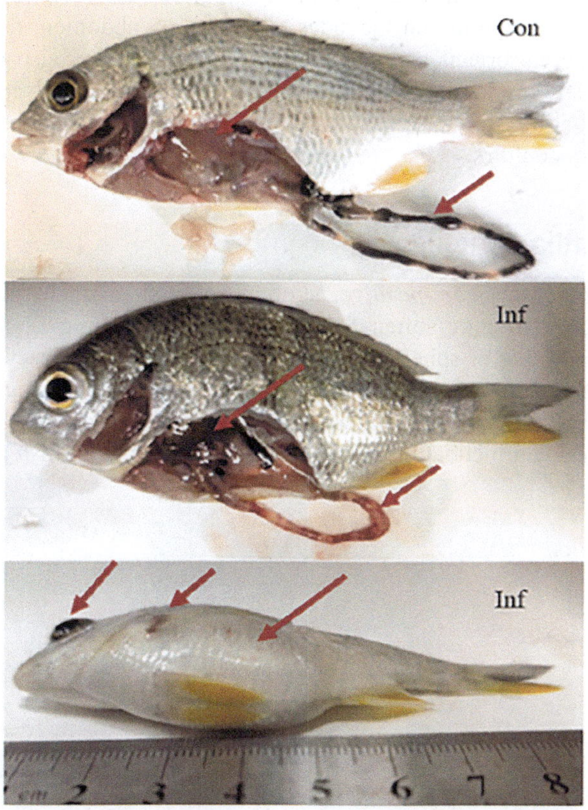

FIGURE 8.5 The fish infected with *Streptococcus agalactis* (Fan et al., 2025).

Shows *Streptococcus agalactiae*-infected fish exhibiting typical clinical signs including systemic hemorrhage, exophthalmia, abdominal effusion, and hemorrhagic lesions at fin bases.

Fan, H-D., Zhao, Y-M., Liu, M., Wang, X., Lin, J., Huang, W., Wang, J-H.,2025. Mechanisms of immune responses in Acanthopagrus latus to Streptococcus agalactiae infection revealed by transcriptomic analysis. Fish & Shellfish Immunology, 157. https://doi.org/10.1016/j.fsi.2025.110114.

25~30°C but can also survive at temperatures as low as 4°C. However, it ceases to grow at temperatures above 41°C. *Pseudomonas fluorescens* is a significant pathogen in aquaculture, responsible for fish erysipelatous dermatitis and acting as a saprophytic bacterium that contributes to the spoilage of aquatic products. This bacterium can infect various small fish species and does not exhibit a clear seasonal epidemic pattern. It primarily affects grass carp and herring but can also infect carp, bream, rainbow trout, mullet, barracuda, flounder, bass, and grouper. The bacteria typically invade fish through wounds in the skin tissue, leading to hemorrhaging, inflammation, erosion, and ulceration on the skin surface (Fig. 8.6). Infected fish commonly show symptoms on the sides of the trunk, abdomen, fins, and gills. In some cases, internal organs, such as the intestinal tract, may also exhibit signs of hemorrhaging and inflammation (Dintürk and Tanrkul, 2021). Given its widespread host range and ability to cause severe infections, *P. fluorescens* poses a significant challenge in aquaculture.

8.2.6 Nocardia seriolae

Nocardia seriolae is a *Gram-positive*, aerobic bacterium characterized by its long or short rod-shaped structure, often forming slender branching filaments that may fragment into rod-like or spherical forms. In recent years, this bacterium has had a significant impact on the aquaculture industry in China, causing nocardiosis in species such as largemouth bass, large yellow croaker, and sea bass. Following infection, there is an incubation period during which affected fish may not exhibit

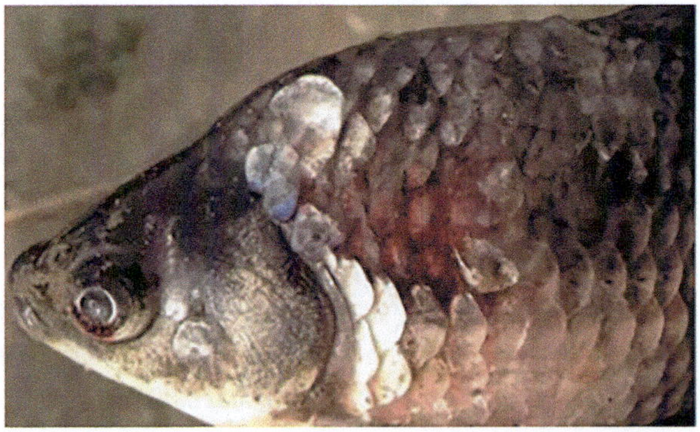

FIGURE 8.6 The fish infected with *Pseudomonas fluorescens* (Baihe County Agriculture and Rural Affairs Bureau, Shanxi, China).

shows *Pseudomonas fluorescens*-infected fish exhibiting characteristic skin lesions including hemorrhage, inflammation, erosion and ulceration, particularly visible on the trunk, abdomen and fins.

any visible external symptoms. Initially, infected fish display subtle behavioral changes, including sluggish movement, reduced appetite, and a tendency to float near the water surface. As the disease progresses, some fish develop darkened skin or white to light-yellow nodules on the body surface, accompanied by ulcerative bleeding. The tail fin may also show signs of ulcerative lesions, eventually leading to mortality. Internally, *N. seriolae* invades multiple organs, causing the formation of white or yellowish nodules in tissues such as the gills, kidneys, liver, spleen, and swim bladder, as illustrated in Fig. 8.7.

8.2.7 Others

Aquaponics is a highly integrated ecosystem that seamlessly connects fish farming and plant cultivation through the activity of microorganisms. Beyond nitrifying bacteria, numerous other beneficial bacteria contribute to maintaining the stability and efficiency of the system. For instance, *Bacillus subtilis* plays a crucial role in suppressing pathogenic bacteria within aquaponics systems, effectively preventing disease outbreaks in both aquatic animals and hydroponic plants. In addition, it aids in decomposing residual feed, fish waste, and organic matter, thereby purifying water quality and enhancing nutrient availability, which significantly boosts fertilizer efficiency and promotes plant growth. Moreover, probiotics within the system include *Lactic acid bacteria, Actinomycetes, Filamentous bacteria, Bifidobacteria, Bacillus species*, and *Photosynthetic bacteria*. These probiotics play a vital role in improving water quality. They also support the healthy growth of

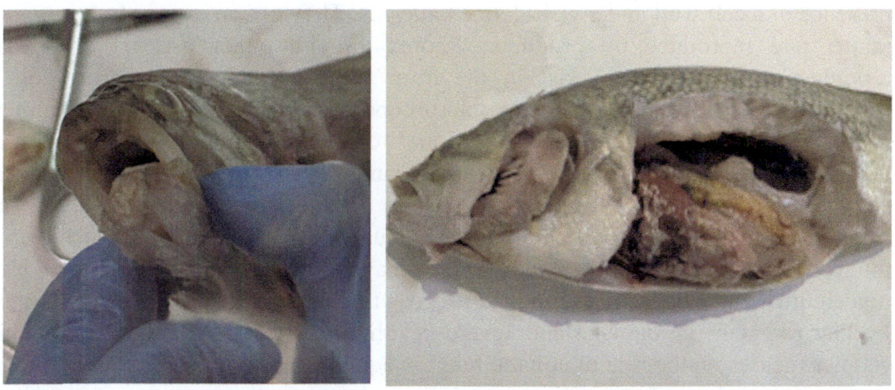

FIGURE 8.7 The fish infected with *Nocardia seriolae* (National Digital Fisheries Innovation Center, China Agricultural University, China).

Shows *Nocardia seriolae*-infected fish exhibiting characteristic white-to-yellow nodules on the body surface and fins, along with internal granulomas in multiple organs.

both hydroponic plants and aquatic animals. In addition, they enhance nutrient absorption and utilization. Furthermore, they strengthen the disease resistance of the system. By fostering a balanced microbial community, these beneficial bacteria contribute to the overall sustainability and productivity of the aquaponics ecosystem.

As aquaculture water is continuously recycled, the accumulation of aquatic animal waste and excrement can contribute to water pollution, significantly impacting the composition and structure of microbial communities. These changes can alter the balance between beneficial and pathogenic bacteria in the system, potentially disrupting the microbial equilibrium. When water quality deteriorates, the increased presence of harmful bacteria may lead to disease outbreaks among aquatic organisms. Beyond the common pathogenic bacteria previously mentioned, aquaponics systems are also susceptible to the introduction of harmful microorganisms through contaminated feed, water sources, seeds, and animal waste. Pathogens such as *Escherichia coli*, *Salmonella*, *Shigella*, and *Staphylococcus aureus* pose serious risks, not only contributing to aquatic animal diseases but also raising concerns about the safety of agricultural and aquatic food products. Therefore, strict monitoring and preventive measures are essential to maintaining water quality, ensuring system stability, and safeguarding both aquatic organisms and food safety.

8.3 Fungi

Fungi are a diverse group of eukaryotic microorganisms characterized by the presence of a cell wall and a well-defined nucleus. They obtain energy from organic matter and reproduce by producing spores through either asexual or sexual reproduction. Most fungi exist in the form of mycelium, which grows on a substrate to form colonies. Under a microscope, fungal hyphae appear as tubular structures. While mycelium has the potential for indefinite growth, its diameter is generally limited to 2~30 μm. In addition to filamentous fungi, some species exist as single-celled organisms, either as oval-shaped *Yeasts* with a cell wall or as plasmodial forms lacking a cell wall. In aquaponics systems, when fish come into contact with fungal spores, these spores may colonize the host, developing into mycelium within the fish. This mycelial growth can then spread to other tissues, leading to extensive damage and severe health issues (Stajich et al., 2009). This chapter focuses on the role of aquatic fungi in aquaponics systems. Beneficial fungi, such as *Yeasts*, contribute to system stability, whereas pathogenic fungi, including *Saprolegnia*, *Branchiomyces*, and *Fusarium*, pose significant threats to aquatic organisms. Fungi play a crucial role in maintaining ecological balance within aquaponics, making their study essential for ensuring the system's stability and long-term sustainability.

8.3.1 Yeasts

Yeasts are a group of unicellular eukaryotic microorganisms that are typically round or oval in shape. They primarily reproduce through budding or binary fission, and more than 1000 species of *Yeast* have been identified to date (Mukherjee and Ghorai, 2022). The cell structure of *Yeast* is illustrated in Fig. 8.8. In laboratory settings, *Yeast* is commonly cultivated using nutrient-rich media composed of potato extract, wort, cornstarch, soybean juice, glucose, sucrose, and *Yeast* extract, among other ingredients. Most *Yeast* species are mesophilic organisms, with an optimal growth temperature typically ranging between 20°C and 30°C. Under aerobic conditions, *Yeast* grows rapidly, while under anaerobic or low-oxygen conditions, it ferments sugars to produce alcohol. This unique metabolic versatility makes *Yeast* an essential microorganism in various biological and industrial applications.

In aquaponics systems, *Yeast* plays several essential roles in maintaining ecological balance and promoting overall system health. *Yeast* is a rich source of high-quality proteins, amino acids, fatty acids, vitamins, and minerals, making it a valuable nutritional supplement for both fish and plants. Beyond its nutritional benefits, *Yeast* produces various enzymes, including cellulase, amylase, and phytase, which help break down organic matter and enhance the activity of other beneficial microorganisms. This enzymatic activity supports nutrient cycling and

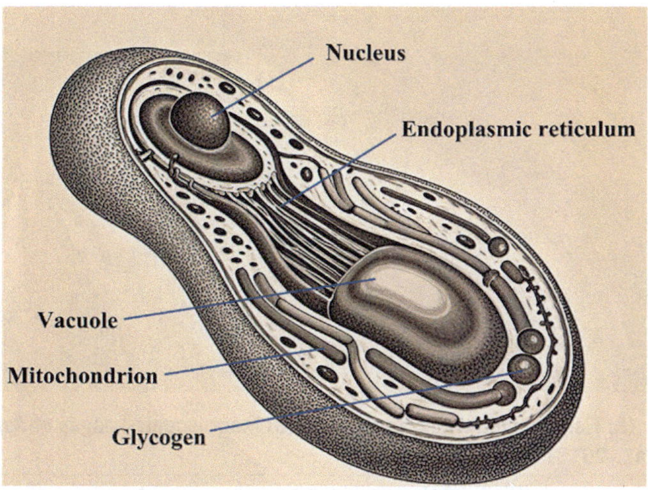

FIGURE 8.8 Yeasts cell structure.

illustrates the typical cell structure of yeast, showing key cellular components of this unicellular eukaryotic microorganism.

contributes to a healthier microbial ecosystem. In addition, *Yeast* serves as a biocontrol agent by occupying ecological niches, thereby limiting the proliferation of harmful bacteria through competitive exclusion (Babbal et al, 2017). This mechanism helps maintain microbial stability and indirectly improves the overall health of the aquaponics system. Furthermore, *Yeast* extracts can be incorporated into fish feed to enhance nutritional intake. As a feed additive, *Yeast* has been shown to promote growth rates and improve fish health, making it a crucial component in sustainable aquaculture practices.

8.3.2 Saprolegniosis

Saprolegnia belongs to the *Kingdom Stramenopila*, *Mastigomycotina*, *Oomycetes*, *Saprolegniales*, and *Saprolegniaceae*. It is a widespread aquatic pathogen capable of infecting almost all freshwater fish species, including tilapia, trout, and grass carp. The flagellated spores of *Saprolegnia* can actively swim in water, searching for suitable fish hosts. Once attached, the fungus develops cotton-like mycelium, as shown in Fig. 8.9. Infection typically occurs through the gills or wounds on the fish's body surface. While healthy fish are generally resistant to *Saprolegnia* infections, fish in poorly managed aquaponics systems or those with injuries are significantly more susceptible. In the early stages, *Saprolegnia* infections are difficult to detect with the naked eye. As the disease progresses, white or yellow cotton-like flocculent mycelium appears on the fish's surface. These fungal

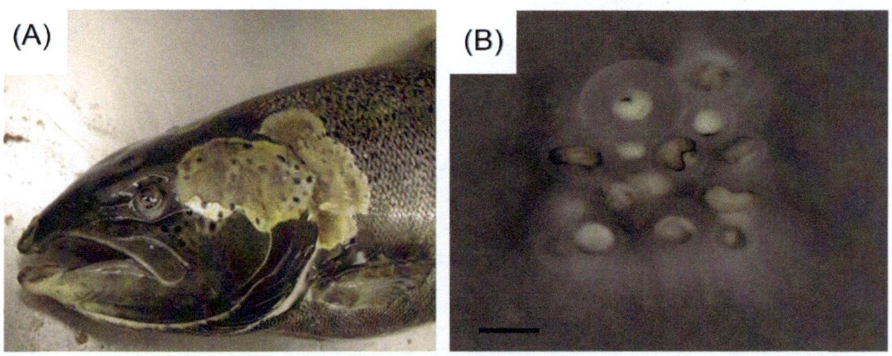

FIGURE 8.9 (A) Fish infected with *Saprolegnia*; (B) Macroscopic images of *Saprolegnia* (van den Berg et al., 2013).

(a) shows fish infected with *Saprolegnia*, exhibiting characteristic white cotton-like mycelial growth on the body surface; (b) presents a macroscopic view of the fungal structures.

van den Berg, A.H., McLaggan, D., Diéguez-Uribeondo, J., van West, P.,2013 The impact of the water moulds Saprolegnia diclina and Saprolegnia parasitica on natural ecosystems and the aquaculture industry. Fungal Biology Reviews, 27(2), 33–42. https://doi.org/10.1016/j.fbr.2013.05.001.

filaments can penetrate deep into tissues, leading to localized ulceration and excessive mucus secretion. Infected fish often exhibit behavioral changes, such as rubbing against surfaces, which can cause the mycelium to detach and disperse into the water (Fig. 8.9), increasing the risk of transmission (van den Berg et al., 2013).

8.3.3 Branchiomyces

Branchiomyces belongs to the *Saprolegniales* and *Saprolegniaceae*. This fungal pathogen is responsible for branchiomycosis, a disease that primarily affects the gills of fish, causing inflammation and, in severe cases, leading to mortality (Khoo, 2000). *Branchiomycosis* predominantly affects freshwater farmed fish, including tilapia, grass carp, common carp, yellow catfish, catfish, and bighead carp. The disease is most prevalent during hot seasons, when environmental conditions favor fungal proliferation. In its early stages, *Branchiomycosis* is difficult to detect with the naked eye. Infection begins when spores attach to the gills, developing into mycelium, which then extends inward and branches further. As the mycelium spreads along the blood vessels in the gills, it can block microvessels, as illustrated in Fig. 8.10 (Mahboub and Shaheen, 2021) This vascular obstruction leads to tissue damage, causing the gill flaps to lose their natural red coloration. As the infection progresses, fish develop swollen gill filaments, experience excessive mucus production, and may even suffer from ulceration. *Branchiomycosis* often manifests as an acute outbreak, with rapid disease progression. Under favorable conditions, *Branchiomyces* can multiply within 1–2 days, leading to mass mortality in

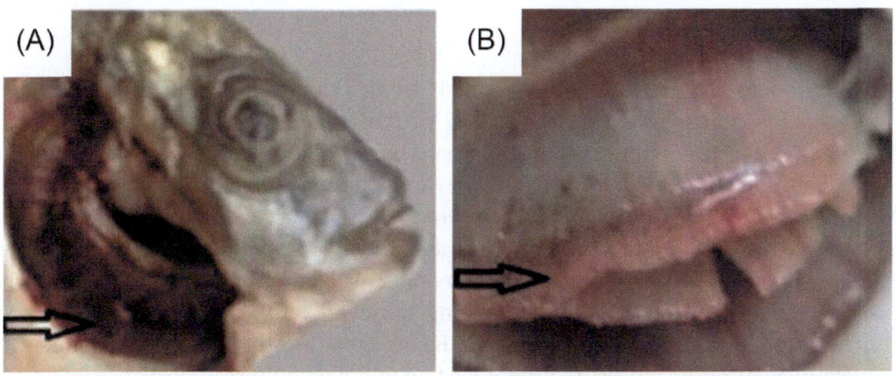

FIGURE 8.10 Fish gills infected with *Branchiomyces* (A,B) (Mahboub and Shaheen, 2021).

Shows fish gills infected with *Branchiomyces*, displaying characteristic tissue damage including vascular obstruction, filament swelling and discoloration caused by fungal mycelium growth.

Mahboub, H.H., shaheen, A.A., 2021. Mycological and histopathological identification of potential fish pathogens in Nile tilapia. Aquaculture, 530. https://doi.org/10.1016/j.aquaculture.2020.735849.

infected ponds, with death rates reaching 60% or higher. Due to its fast-spreading nature, early detection, water quality management, and preventive measures are crucial in controlling branchiomycosis and mitigating losses in aquaponics and aquaculture systems.

8.3.4 Fusarium

Fusarium belongs to the *Ascomycota*, *Sordariomycetes*, *Hypocreales*, and *Nectriaceae*. It is characterized by transparent to light-colored mycelia with well-developed branching structures. When cultured on media, *Fusarium* colonies typically exhibit a villous texture, with colors ranging from white to pink. This fungal genus poses a significant threat to both plants and fish in aquaponics systems (Rossman et al., 2013). In plants, *Fusarium* can cause root and stem rot as well as leaf necrosis, leading to reduced plant health and yield. In fish, the earliest symptom of *Fusarium* infection is inflammation of the body. Once spores penetrate the epidermis, they spread throughout the tissues and organs via the bloodstream. During spore germination, the infection leads to congestion, hemorrhaging, and thrombosis, disrupting blood circulation and causing tissue necrosis. As the disease progresses, infected fish develop ulcerations, and their bodies become covered in white mycelium (as shown in Fig. 8.11) (Cutuli et al., 2015). Internally, the liver, intestines, and kidneys become congested and swollen, and severe infections lead to ulcers and perforations in the abdominal cavity. Left unchecked, Fusarium infections can severely compromise fish health and disrupt the stability of the entire aquaponics ecosystem.

8.3.5 Others

In addition to the previously mentioned fungi, other fungal infections can also impact aquaponics systems, including *Pythium*, *Penicillium*, and *Mucor*. While these fungi have a less direct impact on fish (Dinev et al., 2023), they can spoil feed and plants, leading to deterioration of water quality and indirectly affecting the overall balance of the aquaponics ecosystem. To prevent fungal infections, proactive measures are essential. Key strategies include maintaining optimal water quality, ensuring proper environmental hygiene, selecting healthy fish species, and using high-quality feed. By implementing these preventive measures, the risk of fungal infections can be significantly reduced, ensuring the stable development of aquaponics systems and promoting a sustainable aquaculture industry.

8.4 Viral

Viral infections in fish are a major concern in aquaponics systems. The intensive environment and water circulation system create ideal conditions for virus transmission, especially when water quality management is inadequate or the

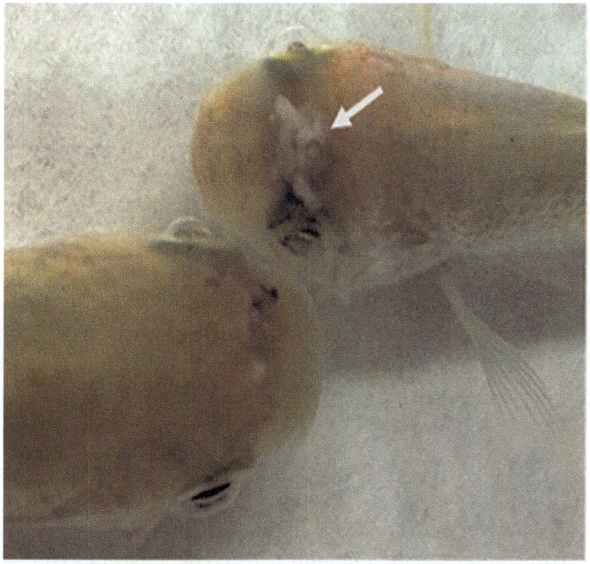

FIGURE 8.11 The symptom of fusarium (Cutuli et al., 2015).

Shows Fusarium-infected fish exhibiting characteristic symptoms including white mycelial growth on the body surface, ulcerative lesions, and internal organ congestion.

Cutuli, M.T., Gibello, A., Rodriguez-Bertos, A., Blanco, M.M., Villarroel, M., Giraldo, A., Guarro, J., 2015. Skin and subcutaneous mycoses in tilapia (Oreochromis niloticus) caused by Fusarium oxysporum in coinfection with Aeromonas hydrophila. Medical Mycology Case Reports, 9, 7–11. https://doi.org/10.1016/j.mmcr.2015.06.002.

feeding environment is unclean. Poor system maintenance increases the likelihood of viral outbreaks, which can lead to severe consequences. Once a virus infects fish within an aquaponics system, the effects can be devastating, potentially causing mass fish mortalities and water quality deterioration. As a result, the organic matter from fish waste decreases. This causes nutrient deficiencies and lowers plant yield. Several common viruses that infect fish in aquaponics systems include *Reovirus*, *Herpesvirus*, *Spring viremia of carp virus*, and *Iridovirus*. Given the potential impact on both fish health and overall system stability, strict biosecurity measures, proper water quality management, and regular health monitoring are essential to prevent and mitigate viral infections in aquaponics systems.

8.4.1 Reovirus

Reovirus belongs to the *Resentoviricetes*, *Reovirales*, and *Reoviridae*. It is an RNA virus that primarily infects grass carp species in freshwater aquaculture (Kibenge, 2024). *Rotavirus*, a type of reovirus, is a spherical virus with a diameter of

70~80 nm. *Reovirus* particles have a smooth surface, consisting of a core (nucleus) and an outer capsid. The core contains the viral genomic RNA, while the outer capsid is composed of multiple structural proteins. These proteins serve a dual function, protecting the viral particles and binding to host cell receptors, facilitating viral entry and infection. In the early stages of infection, infected fish develop black spots on their bodies. They exhibit abnormal swimming behavior, often isolating themselves near the water surface, becoming lethargic, and showing reduced appetite. The primary symptom of reovirus infection is congestion, with spotty or lumpy hemorrhages appearing across various organs and tissues, including the fins, gill covers, gill filaments, eye sockets, mouth, jaws, and other epidermal tissues. Due to these widespread hemorrhagic symptoms, the disease is commonly referred to as "viral hemorrhagic disease". In addition, infected fish often exhibit protruding eyes, further distinguishing the infection (Fig. 8.12) (Xu et al., 2013).

8.4.2 Herpesvirus

Herpesviruses are classified into three major families based on genomic analysis, *Herpesviridae*, *Malacoherpesviridae*, and *Alloherpesviridae* (McGeoch et al, 2006). Among these, *Alloherpesviridae* primarily infect fish and amphibians. It is further

FIGURE 8.12 The symptom of reovirus (Xu et al., 2013).

Shows reovirus-infected fish exhibiting characteristic symptoms of viral hemorrhagic disease, including widespread spot-like hemorrhages on body surfaces and fins, along with protruding eyes.

Xu, J., Zeng, L., Luo, X., Wang, Y., Fan, Y., Gong, S.,2013. Reovirus infection emerged in cultured channel catfish, Ictalurus punctatus, in China. Aquaculture, 372–375, 39–44. https://doi.org/10.1016/j.aquaculture.2012.10.011.

divided into three genera that infect fish, *Cyprinivirus*, *Ictalurivirus*, and *Salmonivirus*. Fish species commonly affected by *Alloherpesviridae* include carp, crucian carp, grass carp, salmon, trout, eel, and salvelinus species. Infected fish often exhibit redness, swelling, and ulceration on the body surface. Behavioral abnormalities such as slow swimming, sinking, or struggling at the water surface may also be observed. In addition, edema or hemorrhaging can occur in the eyes, while fins may appear decayed or damaged, displaying irregular edges or discoloration. In severe cases, internal organs may also suffer from hemorrhaging or necrosis in Fig. 8.13 (Wu et al., 2013).

8.4.3 Spring viremia of carp virus

Spring viremia of carp virus (SVCV) is a serious aquatic animal disease caused by a member of the genus Sprivivirus in the family Rhabdoviridae. SVCV is a typical rhabdovirus with a bullet-shaped morphology, measuring approximately 80~200 nm in length and 50~80 nm in width. It primarily affects cyprinid fish, including species such as *Cyprinus carpio* and *Carassius auratus*. Infected fish commonly exhibit clinical signs such as loss of appetite, respiratory distress, and abnormal swimming behavior, including lethargy and loss of equilibrium. External symptoms may include exophthalmia (protrusion of the eyes), abdominal distension, and redness or swelling of the vent. In severe cases, systemic hemorrhaging may occur, evident as petechiae on the gills, skin, and internal organs such as the swim bladder, liver, spleen, and kidney (as shown in Fig. 8.14) (Zheng et al., 2018).

FIGURE 8.13 The symptom of *herpesvirus* (Wu et al., 2013).

(A) Diseased crucian carp showing gill/fin base ecchymosis and pale fins; (B) Petechial hemorrhage on the swim bladder of diseased crucian carp. Shows herpesvirus-infected fish exhibiting characteristic symptoms including body surface ulceration, fin erosion, ocular edema, and hemorrhaging.

Wu, T., Ding, Z., Ren, M., An, L., Xiao, Z., Liu, P., Gu, W., Meng, Q., Wang, W., 2013. The histo- and ultrapathological studies on a fatal disease of Prussian carp (Carassius gibelio) in mainland China associated with cyprinid herpesvirus 2 (CyHV-2). Aquaculture, 412–413, 8–13. https://doi.org/10.1016/j.aquaculture.2013.07.004

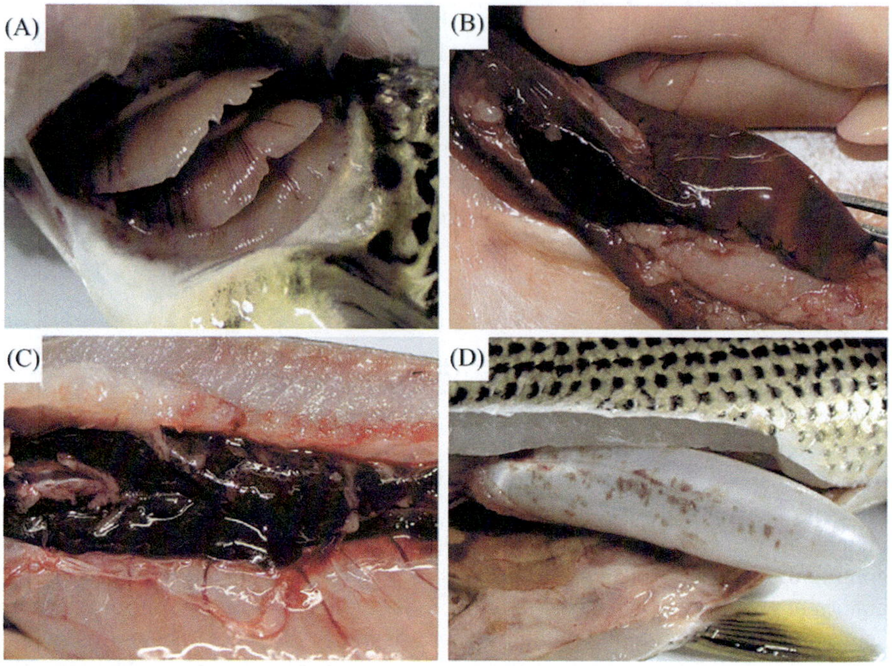

FIGURE 8.14 The symptom of *spring viremia of carp* (Zheng et al., 2018).

(A) Gills pale and hemorrhage; (B) Hepatomegaly and splenomegaly, with hemorrhage; (C) Kidney tumefaction and hemorrhage; (D) Petechial hemorrhages in the wall of the swim bladder. Shows carp infected with spring viremia of carp virus (SVCV), exhibiting typical symptoms including abdominal distension, vent redness, and systemic hemorrhaging on gills and body surface.

Zheng, L.P., Geng, Y., Yu, Z.H., Wang, K.Y., Ou, Y.P., Chen, D.F., Huang, X.L., Deng, L.J., Gan, W.X., Fang, J., Zhong, Z.J., Lai, W.M., 2018. First report of spring viremia of carp virus in Percocypris pingi in China. Aquaculture, 493, 214–218. https://doi.org/10.1016/j.aquaculture.2018.04.056.

8.4.4 Iridovirus

Iridoviruses are a group of DNA viruses with diameters ranging from 120 to 300 nm. Once these viruses infect host cells, they encode and produce various enzymes that inhibit nucleic acid replication and protein synthesis. This disruption interferes with the normal cellular metabolism, ultimately leading to cell death. Additionally, Iridovirus compromises the integrity of the lysosomal membrane, causing lysosomal enzyme leakage and triggering cell autolysis. The incubation period of Iridovirus ranges from 5 to 15 days. During an outbreak, the disease spreads rapidly, with mortality rates reaching up to 100%. These viruses have an

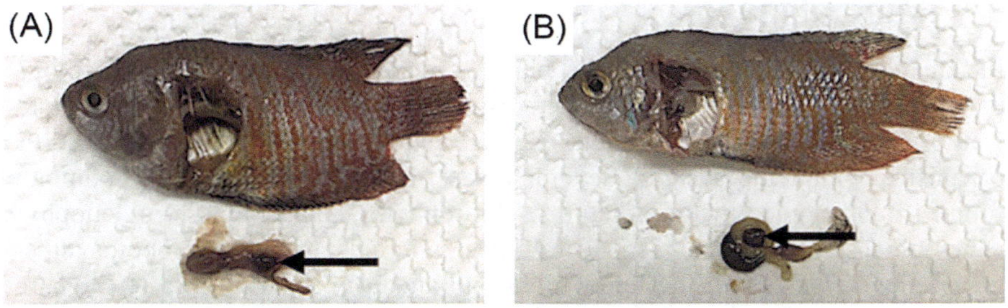

FIGURE 8.15 The symptom of *Iridovirus* (Chuang et al., 2022).

(A) Diseased fish showing splenomegaly; (B) Diseased fish showing splenomegaly with associated tissue discoloration. Shows Iridovirus-infected fish exhibiting characteristic symptoms including darkened body coloration, gill anemia, surface hemorrhage, and visceral organ hypertrophy.

Chuang, H.C., Chu, T.W., Cheng, A.C., Chen, N.Y., Lai, Y.S., 2022. Iridovirus isolated from marine giant sea perch causes infection in freshwater ornamental fish. Aquaculture, 548. https://doi.org/10.1016/j.aquaculture.2021.737588.

extensive host range, infecting nearly 100 aquatic species, including perch and grouper. Infected fish typically exhibit darkened body coloration, gill anemia, and surface hemorrhaging. Internally, the liver, spleen, and kidneys become hypertrophic, as shown in Fig. 8.15 (Chuang et al., 2022). Histological examination of spleen tissue reveals abnormally enlarged cells filled with a high concentration of viral particles. Although vaccines have been developed to combat Iridovirus, their high mutation rate often reduces the effectiveness of vaccination. Given its rapid transmission and high fatality rate, early detection, strict biosecurity measures, and comprehensive disease management are crucial to controlling Iridovirus outbreaks in aquaculture systems.

8.4.5 Others

In addition to the major viruses mentioned above, other viruses such as *Carp Edema Virus* and *Infectious Salmon Anemia Virus* also pose significant threats to aquaculture. Viral infections not only cause disease and mortality in fish but also trigger a cascade of negative effects throughout the aquaponics system. First, mass fish mortalities increase the organic load in the system, leading to a rise in ammonia, nitrite, and nitrate levels, which degrades water quality. These changes can reduce the efficiency of nutrient uptake in plants and, in some cases, even cause plant diseases. In addition, residual pathogens may persist in the system, continuously circulating through the water and increasing the risk of infection for other fish. If left unchecked, these disruptions can ultimately destabilize the entire aquaponics ecosystem.

8.5 Microalgae

In aquaponics systems, microalgae play a crucial role in maintaining ecological balance. They function not only as photosynthetic producers, but also contribute to water quality management, nutrient cycling, and the overall stability and health of organisms within the system. Several types of beneficial microalgae commonly found in aquaponics include diatoms, chlorophytes, phaeophytes, euglenophytes, chrysophytes, and cryptoalgae. Most of these microalgae engage in photosynthesis, generating oxygen for the system while simultaneously absorbing waste and carbon dioxide, thus improving water quality. In addition, some species can serve as organic fertilizers, enriching the nutrient profile of the system. However, excessive microalgal growth can lead to eutrophication, which may disrupt the ecological balance. Despite their benefits, certain algae can also pose challenges to aquaponics systems. For instance, cyanobacteria can sometimes trigger harmful algal blooms, leading to oxygen depletion and further exacerbating eutrophication, which negatively impacts both fish and aquatic plants. Furthermore, some cyanobacteria produce toxins that pose potential threats to ecosystem health and human safety. Similarly, phytoplankton overgrowth can result in increased water turbidity, reducing water transparency and light penetration, thereby inhibiting the growth and nutrient absorption of fish and aquatic plants. To maintain a healthy and stable aquaponics system, it is essential to implement effective management strategies for both beneficial and potentially harmful microalgae. This can be achieved through regular water quality monitoring, assessing the structure and abundance of microalgal populations, and applying physical, biological, and chemical control methods as needed. Additionally, selecting appropriate plant-fish combinations and optimal cultivation strategies can help maximize the benefits of microalgae while minimizing potential negative impacts on the system.

8.6 Detection methods and preventive measures

8.6.1 Detection methods

Traditional methods for detecting pathogenic microorganisms primarily include smear microscopy, biochemical experiments, and cell culture. Smear microscopy is a fundamental microscopic examination technique used to analyze microbial samples (as shown in Fig. 8.16). The process includes several steps. First, smear preparation is performed. Then, the sample is stained and observed under a microscope. Finally, data is recorded and analyzed to examine microbial characteristics in detail. Biochemical experiments play a crucial role in microbiology research, allowing the differentiation of microbial species based on metabolic characteristics, enzyme activity, and other biochemical reactions. This method provides valuable insights into microbial identification and classification. Cell culture technology enables the growth and maintenance of cells in an in vitro

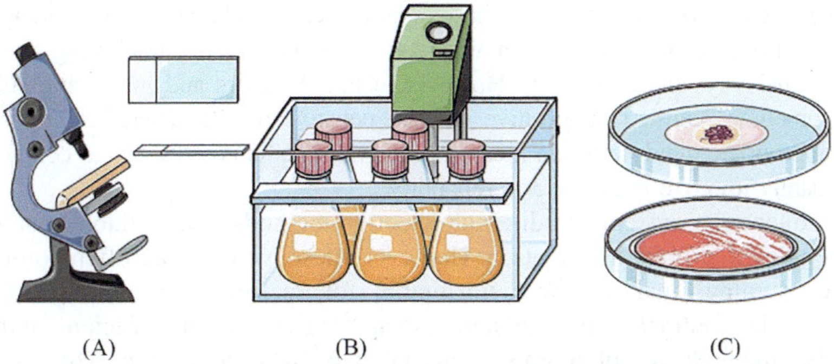

FIGURE 8.16 The traditional detection methods of pathogenic microorganisms.

(A) Microscopic (B) Biochemical experiments (C) Cell culture. Illustrates traditional methods for detecting pathogenic microorganisms, including smear microscopy, biochemical tests, and cell culture techniques used in microbiological analysis.

environment. Through precise techniques such as cell inoculation, cultivation, and observation, researchers can study microbial infection mechanisms, virulence factors, drug resistance, and host-pathogen interactions. However, most traditional detection methods require staining and culturing, which often results in longer detection times, a limited detection range, and lower sensitivity compared to modern diagnostic techniques.

Microbial rapid detection technology is designed to quickly detect and identify microorganisms, providing rapid diagnosis and decision support. The primary rapid detection methods include nucleic acid-based techniques, immunological assays, and biosensors. Nucleic acid detection identifies and quantifies microorganisms by targeting their genetic material (DNA or RNA). Common methods such as polymerase chain reaction (PCR), real-time fluorescence quantitative PCR, and nucleic acid amplification technology offer high sensitivity and specificity, enabling the detection of low concentrations of pathogens within a few hours. Immunological methods detect microbial-specific antigens to identify pathogens. Techniques like enzyme-linked immunosorbent assay, lateral flow assay, and immunofluorescence provide reliable pathogen detection. ELISA utilizes antigen-antibody binding combined with enzymatic reactions to enhance signal amplification for highly sensitive quantitative analysis. LFA is simple to operate, making it ideal for on-site rapid screening, with results available within 15–30 minutes. Immunofluorescence uses fluorescent labeling to detect antigen-antibody interactions, offering high-resolution microbial localization. Biosensor technology, another essential approach, consists of bioreceptors (such as antibodies, aptamers, and enzymes) that specifically bind to microorganisms and signal converters that transform these interactions into detectable output signals, including optical, electrical, or thermal signals. The greatest advantage of biosensors is their capability for real-time, portable, and multi-pathogen detection, making them valuable for onsite production monitoring and biological laboratory analysis. These rapid detection technologies are widely applied in medical diagnostics, environmental safety, and public health monitoring due to their speed, sensitivity, and specificity. However, external factors and interfering substances may sometimes cause false-positive or false-negative results. Therefore, selecting the most appropriate rapid detection method based on specific environmental conditions can significantly improve accuracy and reliability.

In aquaponics systems, traditional detection methods offer reliable baseline data and detailed microbial information. However, these methods are often complex and time-consuming, making them unsuitable for real-time monitoring and rapid response. In contrast, rapid detection technologies provide efficient and highly sensitive methods for obtaining microbial data in a short time. This allows for timely adjustments in system management, ensuring the healthy operation and stability of the aquaponics system. Looking ahead, the integration of various detection technologies with intelligent monitoring equipment will further enhance management efficiency and system productivity. This technological advancement will provide strong support for the sustainable development of aquaponics systems, optimizing both fish farming and plant cultivation within the symbiotic ecosystem.

8.6.2 Preventive measures

Currently, the prevention of aquatic animal diseases remains the primary approach, with a focus on controlling infectivity and developing targeted vaccines to enhance fish immunity. Effective prevention and control strategies center on maintaining optimal water quality parameters, such as temperature, pH levels, and oxygen content, to limit the survival and reproduction of harmful microorganisms. In addition, promoting biodiversity within the system, particularly by increasing the abundance and diversity of beneficial microorganisms, strengthens the system's resistance to pathogens. Regular water quality monitoring and microbiological testing during production are essential for early detection and identification of potential microbial threats. Implementing preventive measures and emergency response plans is crucial for rapidly addressing disease outbreaks and controlling the spread of harmful microorganisms. Moreover, strict biosecurity measures should be enforced, including quarantine and isolation of new species, as well as disinfection of equipment and materials, to prevent the introduction and transmission of pathogens. By comprehensively applying these strategies, aquaponics systems can effectively manage and control harmful microorganisms, ensuring system stability, ecological health, and high production efficiency.

References

Armwood, A.R., Griffin, M.J., Richardson, B.M., Wise, D.J., Ware, C., Camus, A.C., 2022. Pathology and virulence of *Edwardsiella tarda*, *Edwardsiella piscicida*, and

Edwardsiella anguillarum in channel (*Ictalurus punctatus*), blue (Ictalurus furcatus), and channel × blue hybrid catfish. Journal of Fish Diseases 45 (11), 1683–1698. Available at https://doi.org/10.1111/jfd.13691.

Babbal, Adivitiya, Khasa, Y.P., 2017. Microbes as biocontrol agents. Probiotics and Plant Health. Springer Singapore, India, pp. 507–552.

Chuang, H.C., Chu, T.W., Cheng, A.C., Chen, N.Y., Lai, Y.S., 2022. Iridovirus isolated from marine giant sea perch causes infection in freshwater ornamental fish. Aquaculture (Amsterdam, Netherlands) 548. Available at https://doi.org/10.1016/j.aquaculture.2021.737588.

Cutuli, M.T., Gibello, A., Rodriguez-Bertos, A., Blanco, M.M., Villarroel, M., Giraldo, A., et al., 2015. Skin and subcutaneous mycoses in tilapia (*Oreochromis niloticus*) caused by *Fusarium oxysporum* in coinfection with *Aeromonas hydrophila*. Medical Mycology Case Reports 9, 7–11. Available at https://doi.org/10.1016/j.mmcr.2015.06.002.

Dinev, T., Velichkova, K., Stoyanova, A., Sirakov, I., 2023. Microbial pathogens in aquaponics potentially hazardous for human health. Microorganisms 11 (12), 2824. Available at https://doi.org/10.3390/microorganisms11122824.

Dintürk, Z., Tanrkul, T.T., 2021. *Yersinia ruckeri* and *Pseudomonas fluorescens* co-infection in rainbow trout (Oncorhynchus mykiss Walbaum, 1792). Aquaculture Research 52, 4858–4866. Available at https://doi.org/10.1111/are.15320.

Dong, H.T., Techatanakitarnan, C., Jindakittikul, P., Thaiprayoon, A., Taengphu, S., Charoensapsri, W., et al., 2017. *Aeromonas jandaei* and *Aeromonas veronii* caused disease and mortality in Nile tilapia, *Oreochromis niloticus* (L). Journal of Fish Diseases 40 (10), 1395–1403. Available at https://doi.org/10.1111/jfd.12617.

Fan, H.-D., Zhao, Y.-M., Liu, M., Wang, X., Lin, J., Huang, W., et al., 2025. Mechanisms of immune responses in *Acanthopagrus latus* to *Streptococcus agalactiae* infection revealed by transcriptomic analysis. Fish & Shellfish Immunology 157, 110114. Available at https://doi.org/10.1016/j.fsi.2025.110114.

Haenen, O.L.M., Dong, H.T., Hoai, T.D., Crumlish, M., Karunasagar, I., Barkham, T., et al., 2023. Bacterial diseases of tilapia, their zoonotic potential and risk of antimicrobial resistance. Reviews in Aquaculture 15 (S1), 154–185. Available at https://doi.org/10.1111/raq.12743.

Kasozi, N., Abraham, B., Kaiser, H., Wilhelmi, B., 2021. The complex microbiome in aquaponics: significance of the bacterial ecosystem. Annals of Microbiology 71 (1). Available at https://doi.org/10.1186/s13213-020-01613-5.

Khoo, L., 2000. Fungal diseases in fish. Seminars in Avian and Exotic Pet Medicine 9 (2), 102–111. Available at https://doi.org/10.1053/ax.2000.4623.

Kibenge, F.S.B., 2024. Chapter 1 - Classification and identification of aquatic animal viruses. In: Kibenge, F.S.B., Godoy, M.G. (Eds.), Aquaculture Virology. Academic Press, pp. 3–47. Available at https://doi.org/10.1016/B978-0-323-91169-6.00018-2.

Mahboub, H.H., shaheen, A.A., 2021. Mycological and histopathological identification of potential fish pathogens in Nile tilapia. Aquaculture (Amsterdam, Netherlands) 530, 735849. Available at https://doi.org/10.1016/j.aquaculture.2020.735849.

McGeoch, D.J., Rixon, F.J., Davison, A.J., 2006. Topics in herpesvirus genomics and evolution. Virus Research 117 (1), 90–104. Available at https://doi.org/10.1016/j.virusres.2006.01.002.

Mukherjee, S., Ghorai, S., 2022. 3 - Fungal biology. In: Taherzadeh, M.J., Ferreira, J.A., Pandey, A. (Eds.), Fungal biology Current Developments in Biotechnology and

Bioengineering: Filamentous Fungi Biorefinery. Elsevier, pp. 67–104. Available at https://doi.org/10.1016/B978-0-323-91872-5.00017-X.

Rossman, A.Y., Seifert, K.A., Samuels, G.J., Minnis, A.M., Schroers, H.J., Lombard, L., et al., 2013. Genera in Bionectriaceae, Hypocreaceae, and Nectriaceae (Hypocreales) proposed for acceptance or rejection. IMA Fungus 4 (1), 41–51. Available at https://doi.org/10.5598/imafungus.2013.04.01.05.

Stajich, J.E., Berbee, M.L., Blackwell, M., Hibbett, D.S., James, T.Y., Spatafora, J.W., et al., 2009. The fungi. Current Biology 19 (18), R840. Available at https://doi.org/10.1016/j.cub.2009.07.004.

van den Berg, A.H., McLaggan, D., Diéguez-Uribeondo, J., van West, P., 2013. The impact of the water moulds *Saprolegnia diclina* and *Saprolegnia parasitica* on natural ecosystems and the aquaculture industry. Fungal Biology Reviews 27 (2), 33–42. Available at https://doi.org/10.1016/j.fbr.2013.05.001.

Wu, T., Ding, Z., Ren, M., An, L., Xiao, Z., Liu, P., et al., 2013. The histo- and ultra-pathological studies on a fatal disease of Prussian carp (*Carassius gibelio*) in mainland China associated with cyprinid herpesvirus 2 (CyHV-2). Aquaculture (Amsterdam, Netherlands) 412-413, 8–13. Available at https://doi.org/10.1016/j.aquaculture.2013.07.004.

Xu, J., Zeng, L., Luo, X., Wang, Y., Fan, Y., Gong, S., 2013. Reovirus infection emerged in cultured channel catfish, *Ictalurus punctatus*, in China. Aquaculture (Amsterdam, Netherlands) 372-375, 39–44. Available at https://doi.org/10.1016/j.aquaculture.2012.10.011.

Zheng, L.P., Geng, Y., Yu, Z.H., Wang, K.Y., Ou, Y.P., Chen, D.F., et al., 2018. First report of spring viremia of carp virus in *Percocypris pingi* in China. Aquaculture (Amsterdam, Netherlands) 493, 214–218. Available at https://doi.org/10.1016/j.aquaculture.2018.04.056.

Zou, Y., Hu, Z., Zhang, J., Xie, H., Liang, S., Wang, J., et al., 2016. Attempts to improve nitrogen utilization efficiency of aquaponics through nitrifies addition and filler gradation. Environmental Science and Pollution Research 23 (7), 6671–6679. Available at https://doi.org/10.1007/s11356-015-5898-0.

CHAPTER

Smart aquaponics farming integration

9

Daoliang Li[1,2], Liang Wang[1,3] and Xianbao Xu[2,4]

[1]*National Innovation Center for Digital Fishery, China Agricultural University, Beijing, P.R. China,*
[2]*College of Information and Electrical Engineering, China Agricultural University, Beijing, P.R. China,*
[3]*School of Chemistry and Chemical Engineering, University of Surrey, Guildford, United Kingdom,*
[4]*College of Electronics and Electrical Engineering, Anhui Agricultural University, Hefei, Anhui, P.R. China*

9.1 Introduction

Aquaponics focuses on building an online environmental monitoring system, configuring water quality testing, weather stations, video surveillance and other monitoring equipment, realizing real-time monitoring of water quality and the weather environment (Li et al., 2020); informatization transformation of farms, configuring equipment for circulating water treatment and control, automatic and accurate feeding of bait, monitoring and early warning of aquaponics diseases, quality and safety traceability, and supporting the establishment of a system that covers online monitoring, production management and comprehensive management and security; establishing an intelligent system for aquaponics to realize digital and intelligent management of production and control of product quality. The integrated structure of aquaponics system is shown in the Fig. 9.1, which is centered around the infrastructure of aquaponics factory to carry out data collection, transmission, processing and control, and the whole process is automated, and the core of which is the data processing and intelligent decision-making of the cloud platform.

9.2 Measurement and control system integration

Measurement and control system play an important role in aquaponics system, which is the foundation for realizing the intelligence of aquaponics system. Aquaponics supports various network communication methods such as NB-IOT, 4G/5G, TCP/IP, and so on. The data is reported to the center of the cloud platform, so that it can remotely check the on-site production environment, view the historical

Smart Aquaponic Farming. DOI: https://doi.org/10.1016/B978-0-443-28945-3.00011-6

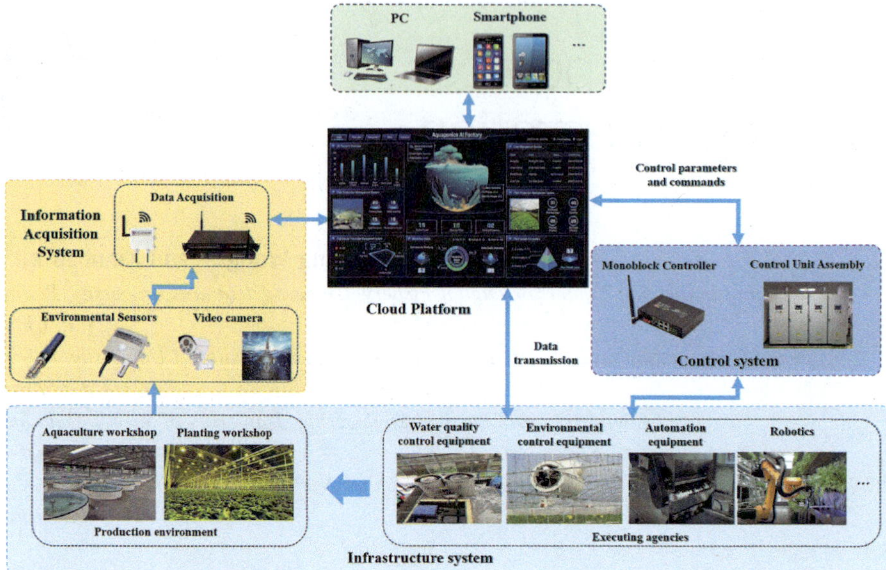

FIGURE 9.1

The integrated structure of aquaponics system.

change curves, analyze the data, and remotely control the operation of the equipment. Deploy data collection equipment such as water quality sensors and high-definition cameras at the breeding site, and transmit the collected data to the data management center through various network transmission methods to carry out real-time monitoring and intelligent control of the breeding production, realizing the functions of early warning of abnormal behaviors, diagnosis of diseases, isolation of faults, quality traceability, and so on; and upload the data to the cloud platform through the shared exchange interface.

9.2.1 Information collection

9.2.1.1 Environmental information

Aquaponics environmental information acquisition is used to monitor and collect environmental data in the aquaponics system, which can help managers understand the water body and environmental conditions, and adjust the culture and planting conditions in time to keep the fish and plants in a suitable growing environment. These information are collected in real time by on-site sensors, while information that cannot be collected in real time by on-site sensors, such as microbial information of nitrifying bacteria, hydrophilic *Aeromonas*, and the chlorophyll content of vegetation, need to be acquired by manual sampling on a regular basis, which can be obtained through chemical analysis or hand-held instrumentation, and thus need to be manually input into the corresponding data management unit of the cloud platform.

9.2.1.1.1 Aquaculture water quality

Aquaculture water quality determines the growth of fish in an aquaponics system. Water quality monitoring reads various parameters in real time through sensors, including pH, dissolved oxygen, temperature, conductivity, and ammonia nitrogen (Somerville et al., 2014). These sensors are generally installed in the middle of the water column or at the flow to measure the content of each parameter in the water, and temperature data are generally obtained from dissolved oxygen sensors without the need for a separate sensor. Water-level sensors are also used in aquaculture pond environments and are generally installed at the bottom of the pond to monitor water level height due to the pressure transducer principle. Data acquisition from these sensors requires the use of appropriate data acquisition equipment, which reads and converts the signals from the sensors and sends the data from each sensor to a local terminal and to the cloud. Data acquisition devices usually use RS485 communication to obtain sensor data. RS485 adopts differential signal transmission and supports multiple devices to communicate through the same signal line, which is suitable for networking applications and commonly used for data communication in long-distance and high-noise environments. In order to reduce the limitation of terrain and wiring, the data collector adopts wireless transmission to send data, which can be accessed to the local wireless network or directly transmitted to the cloud through the mobile communication network, and the local device accesses the collector's data through the cloud or the network IP, so as to obtain the sensor's data information.

9.2.1.1.2 Greenhouse environment

Greenhouse environment has an extremely important impact on aquaponics, and greenhouse environment monitoring mainly includes air temperature and humidity, CO_2 concentration, light intensity, etc. These sensors can provide data basis for air environment regulation in aquaponics system. Sensor selection for greenhouse air temperature and humidity, CO_2 concentration and light intensity, the deployment needs to be based on the layout of the greenhouse and the needs of the deployment of sensors in different locations in order to obtain comprehensive data, to ensure that the sensor is not directly affected by lighting, water, or other interference factors, and can accurately reflect the environmental parameters of the greenhouse. Temperature sensors can be selected based on thermal elements, while humidity sensors can be selected capacitive or resistive sensors, temperature, and humidity sensors need to be evenly distributed in all areas of the room, and to avoid direct exposure to the light source to avoid interference. CO_2 sensors use non-dispersive infrared (NDIR) sensors and are evenly distributed in the center and around the greenhouse to obtain accurate data on overall CO_2 concentration. The sensors for light intensity use photoresistor or photodiode sensors, which are placed in representative locations such as near the roof, while needing to avoid being obscured by tree shadows or other obstacles and ensuring that they match the height of the plants in the greenhouse. The greenhouse environment uses wireless to collect sensor data using Wi-Fi or LoRa wireless communication, and the data collector node is located in the middle position of the room to facilitate the reception of sensor signals from all parts of the room, and the data collector node centrally receives and processes the data from each sensor, and achieves local networking and remote uploading of data through fiber-optic access to switches or routing terminals.

9.2.1.2 Plant information

Plant information assesses plant growth by capturing the texture, color, morphology, and elemental content of plants to better regulate their nutritional ratios. Plant information acquisition relies mainly on spectral sensors and vision cameras. Spectral sensors are selected based on the spectral range to be monitored and the application requirements, and it is generally sufficient to install a near-infrared sensor, while vision cameras use ordinary color cameras, which are generally mounted on a fixed stand or a mobile machine that can cover the planting area. The data collected by the spectral sensors and cameras are sent to the local gateway device via network fiber optics, and then sent to the cloud platform via the gateway for further data analysis in the background.

9.2.1.3 Fish information

Fish information relies on visual, sonar, and acoustic technologies to provide data support for fish behavior, culture volume, and morphology. Visual information is carried out in a combination of aquatic and underwater, both cameras use ordinary visible light color camera type, aquatic camera is installed above the aquaculture pool, which can cover the whole aquaculture pool water surface area, and the underwater camera is generally placed in the central part of the water body, and has the lens option function, which can be adjusted according to the actual needs of the camera angle. The camera mounted on the automatic feeding robot can be equipped with infrared camera function, which can assist the visible light camera to obtain more comprehensive information. The sonar sensor is used to collect the distribution of fish, which is installed in the middle and lower part of the breeding tank, and the detection area can cover most of the area of the breeding tank. The acoustic data were acquired using a hydrophone (frequency range: 1–100 kHz), which was installed in the feeding area for better reception of the acoustic signals. The vision camera, sonar and hydrophone devices are connected (Mahon et al., 2022) to the local data acquisition equipment using network fiber optics, which can be retained in the local recording terminal and uploaded to the cloud platform through a gateway for further data analysis and processing.

9.2.1.4 *Equipment information*

The equipment information in the aquaponics system mainly includes motor or pump information (vibration signal, voltage signal, current signal), position information, and equipment or robot status information (Mobley, 2002). The vibration signal of the motor or pump is obtained by the vibration sensor, which generally adopts analog voltage signal output and needs to be connected to the local controller to realize the digitalization of the signal. Voltage and current signals are obtained from power meters, which also provide power information for energy consumption calculation. The meter is generally installed in each control cabinet, divided into total meter and sub-meter, for water pumps, oxygen generators and other high energy-consuming equipment needs to be assigned a separate meter. The meter adopts RS485 communication mode, and the data is forwarded to the data platform by the local controller or collector. Location information is provided by GPS sensors, which are used to realize the positioning of the equipment, usually installed on mobile devices, such as transport vehicles, robots, and so on. This location information needs to be collected by the collector to realize the wireless upload of data. Device status information is provided differently depending on the device, which comes from the sensors configured in the device itself, and the transmission of the information relies on the digital interface provided by the device, such as a serial port or RS485.

9.2.2 Data transmission

Data transmission system integration refers to the integration of multiple independent data transmission systems into a unified system for seamless data transmission and exchange. As different devices may use different communication protocols and data formats, in order to achieve interoperability between systems, it is necessary to define standardized interface specifications to ensure that the various systems can properly exchange data, in which the middleware or gateway device plays the role of a bridge, responsible for connecting the different data transmission systems and realizing the forwarding and conversion of data.

The data interface in the aquaponics system covers sensors, data acquisition devices, control devices, communication devices, and control panels, and so on. Its data interface and communication protocols include USB interface, RS485 interface, and Modbus protocol, Ethernet interface and TCP/IP protocol, Wi-Fi interface and HTTP protocol in order to, the wireless sensing network interface and LoRa protocol, NFC, and so on. The data transmission interface and communication protocol can effectively realize the data exchange and communication between various devices in the aquaponics system, realize the monitoring and control of environmental parameters such as water quality, temperature, and humidity, and guarantee the normal operation of the system.

Sensors, data collectors and controllers are generally used between the RS485 interface and Modbus RTU protocol, Modbus RTU is a serial communication protocol, the use of binary mode of data transmission, which can be easily realized between sensors, data collectors, controllers and other devices for data exchange and control (Yanes et al., 2020). For some sensors that do not have digital communication functions, such as vibration sensors and rotary encoders, these sensors are often analog signal output, so they can be directly accessed to the collector or control system to achieve data conversion within it, and as an output to be sent to the PTZ or for use by the control device. Drive controller is recommended to use the RS485 interface, which has the characteristics of stability and simplicity, can reduce the load of the micro-control system, while facilitating the maintenance or replacement of late.

Data acquisition devices, communication devices, and control panels generally use Ethernet interfaces with the TCP/IP protocol. The Ethernet interface connects the devices through Ethernet cables and realizes data transmission between devices through the TCP/IP protocol stack, which is suitable for the communication of devices in the LAN with fast and stable speed. In order to realize the data transmission between the cameras, data collectors and other devices in the LAN and the cloud platform, network penetration can be realized by setting up the router gateway to achieve two-way interoperability between the LAN and the Internet.

Wi-Fi interface with HTTP protocol is used for wireless transmission of devices, such as data collectors and data centers, automated robots, and control centers. These devices are connected through a wireless LAN and can transmit data using the HTTP protocol for data transmission and remote monitoring between the devices and the Internet (Kyaw and Ng, 2017).

The wireless sensor network interface uses LoRa protocol for communication, which is a low-power, short-range communication wireless sensor network protocol suitable for data transmission and communication of large-scale sensor nodes. The construction of indoor wireless sensor network can be realized with indoor sensors as slave devices and data collection network nodes as master devices, and the network nodes can send data to the control center or cloud platform through the network interface. Local control cabinets are recommended to use LoRa to realize data transmission between control cabinets, through the establishment of the control cabinet master, the information will be summarized and realized with the cloud platform to send and receive data management.

NFC is a short-range wireless communication technology used for near-field communication and data exchange. The fixed-point positioning function of the robot can be realized by sticking the NFC tag at a suitable position on the track.

9.2.3 Control system

The aquaponics control system uses PLC as the controller of each control cabinet, which mainly receives control commands to drive the actuators to work. The actuators of the aquaponics control system include water pumps, valves, automatic irrigators, light control devices, heaters and coolers, and oxygen supply devices.

The water pump is used to circulate, transport, and provide water; the valve is used to regulate the flow and pressure of liquid or gas, the in and out of water flow, and the branching of the circulating water circuit, in order to realize the control of the flow and distribution of the water body; the automatic irrigator is used to automatically apply fertilizer to the plants and irrigate them; the light control device is used to regulate the intensity of the light and the time of light, which is mainly an LED light; the heater and cooler are used to regulate the temperature of the water; the oxygen supply device is used to provide oxygen in the water body; the oxygen supply device is used to provide oxygen in the water body. The devices are used to provide oxygen in the water body, including roots fans, oxygen generators, nano aerators, and so on. These control actuator types play an important role in the aquaponics system by controlling and regulating the water quality, temperature, light, and other factors in order to realize a good growing environment for fish and plants (Wei et al., 2019).

The aquaponics control system adopts both local configuration and remote control to regulate and manage the system. The local control cabinet is equipped with a control panel, which can configure the parameters of each actuator unit, and the remote end sends control parameters and instructions to the controller host through the cloud platform, which receives the parameters and instructions and assigns them to each subcontroller unit, and the controllers receive the parameters and instructions and execute the corresponding actions (Goddek et al., 2015). The aquaponics control system mainly carries out the precise regulation of water quality, indoor temperature and humidity control, light control, CO_2 control and disaster preparedness protection and other aspects of work.

9.2.3.1 Precise water quality control

The circulating water system in the aquaponics system is responsible for sending water from the fish tank or breeding pond to the plant growing area and back, realizing water recycling. Water quality is related to the growth and development of the fish in the aquaponics system. The local controller obtains the circulating water control parameters from the cloud platform, including the control parameters of the circulating water pumps, valves, filtration equipment, oxygen concentrators, oxygen boosters, and so on, and then drives the corresponding actuators to work, so as to realize the regulation and control of the main parameters such as the water temperature and the dissolved oxygen in the breeding area, in order to meet the needs of the breeding object's growth and development. The local control can also obtain the water quality sensor parameter information directly from the cloud platform, and calculate the control parameters of the dependent actuators to drive the equipment operation.

9.2.3.2 Greenhouse temperature and humidity control

The control of room temperature and humidity is crucial for the growth and development of plants. Indoor sensors send data to the head or data center for processing, and the processed environmental equipment regulation parameters are handed over to local controllers, which drive the actuators to work, including ventilation fans, heating and sprinkler equipment, and so on, so as to realize the automatic regulation of temperature and humidity. When the temperature is too high by adjusting the ventilation equipment of the greenhouse, such as windows, ventilation fans or automatic ventilation system, control the flow of indoor and outdoor air to reduce the temperature in the greenhouse, when the need to add heat using heating equipment such as electric heating, gas, and geothermal heat, to provide additional heat, so that the greenhouse to maintain the appropriate temperature. High humidity can be reduced by increasing ventilation, allowing moist air to leave the greenhouse and introducing dry fresh air. For higher humidity plants or specific growth stages, if humidity is insufficient water misting equipment or sprinkler systems need to be utilized to mist the inside of the greenhouse to increase the humidity in the air. Humidification equipment, such as steam humidifiers or ultrasonic humidifiers, can also be used to release water vapor into the interior of the greenhouse to increase the humidity in the air.

9.2.3.3 *Light intensity control*

In aquaponics systems, light is necessary for plant growth. The light control module is used to control the brightness, time, and period of the light source to meet the light needs of plants at different stages. It can automatically adjust the light according to the set parameters or provide feedback control through external input signals (e.g., photosensitive sensors). The environmental light condition is collected by the light sensor, and the regulation parameters are sent down from the local controller or cloud platform. The light control in the aquaponics system adopts a combination of natural light control and artificial light control. In the daytime when the weather is good, natural light is used as the main light source, and the light intensity is controlled by adjusting the sunshade or shade cloth, and this method is simultaneously adjusted dynamically according to the planted crop, season, and time. In the evening or under insufficient light conditions, LED lights, fluorescent lamps artificial light source, to provide light, it is the main dimming method in aquaponics system. The artificial light source can automatically adjust the brightness and color of the light source according to the preset threshold, thus ensuring stable and consistent light conditions and improving production efficiency. As for certain plants and fish, by controlling the duration and interval of light, the photoperiod of different seasons can be simulated to stimulate plant flowering and fish spawning.

9.2.3.4 *CO_2 control*

CO_2 control is designed to provide plants with the proper concentration of CO_2 for photosynthesis. The controller obtains the CO_2 regulation parameters and controls the ventilation or CO_2 supply system. When the concentration falls below the set value, the CO_2 generator can be automatically turned on for replenishment to maintain the proper CO_2 concentration, and the ventilation fan and indoor circulation fan can be turned on to speed up the indoor and outdoor air circulation and distribute the CO_2 evenly to the plant area. The CO_2 regulation parameters are calculated by the cloud platform based on a combination of sensor data, type and number of plants grown, room area, and supply rate.

9.2.3.5 Disaster preparedness protection

Disaster preparedness protection systems are used to respond to unforeseen situations and disasters to safeguard system operations and biological safety. The disaster preparedness protection system includes water quality monitoring and treatment, temperature, and humidity control, emergency oxygen supply, fire and leakage alarms, and backup power supply. When an abnormal failure occurs, the controller sets off alarms or buzzers even though, and sets the blinking and buzzing frequency according to the urgency of the abnormal situation, the more urgent, the higher the frequency. At the same time, the alarm information will be uploaded to the cloud platform monitoring center through the network, and send alarm push to the corresponding staff to remind the staff to deal with it in time. The disaster protection system can play the role of early warning, regulation and protection in the aquaponics system to ensure that the system can run continuously and stably.

Water quality monitoring and treatment: Water quality monitoring and treatment is including water quality sensors and monitoring equipment, can real-time monitoring of pH, dissolved oxygen, ammonia nitrogen, and other indicators in the water body. When the water quality is abnormal or exceeds the safety range, the automatic alarm, if the staff did not deal with in time then take automatic cleaning, replacement of water body operation.

Temperature and humidity control: Temperature and humidity control is through the temperature and humidity sensors to obtain data, if the staff did not deal with it in time then the local control system according to the set thresholds for automatic heating, cooling, and humidity control, to ensure that the environment temperature and humidity in the range suitable for the survival of fish and plants.

Emergency oxygen supply: When there is insufficient oxygen the system sends out an alarm message in time and turns on the oxygen boosting system. In case of emergency, such as power outage or ventilation failure, the backup oxygen supply system is activated to ensure that there is enough oxygen supply in the breeding pool or plant area to maintain the normal respiration of the organisms.

Fire and water leakage alarms: Fire and water leakage alarms are installed and connected to an automatic sprinkler system or an emergency shutdown device to detect fires or water leaks in a timely manner and take emergency measures.

Backup power supply: Configure backup power generation equipment or battery reserve system to activate the backup power supply system in case of power failure to maintain the necessary power supply to ensure the normal operation of the system.

9.3 Cloud management and control platform system integration

The cloud platform is the brain of the aquaponics system, in addition to providing basic cloud service business, its core is to provide intelligent analysis and decision support for the aquaponics system (Karimanzira and Jacobi, 2021). Information collection sensors and devices provide video data, water quality data, greenhouse environment data, growth data, and feeding data for the cloud platform. The management system provides related business in the aquaculture process, such as water quality control, feeding management, disease warning, production management, and quality traceability. Based on the basic data, business data and video data collected by each system, the cloud platform conducts analysis and mining using data models, and visualizes the analysis results to provide decision-making and management support for the aquaponics system. The functional service framework of the cloud platform and control system is shown in Fig. 9.2.

9.3.1 Data system

The data system is used to store, process, and manage the data in the cloud platform, and the data is transmitted from various sources to the cloud platform through

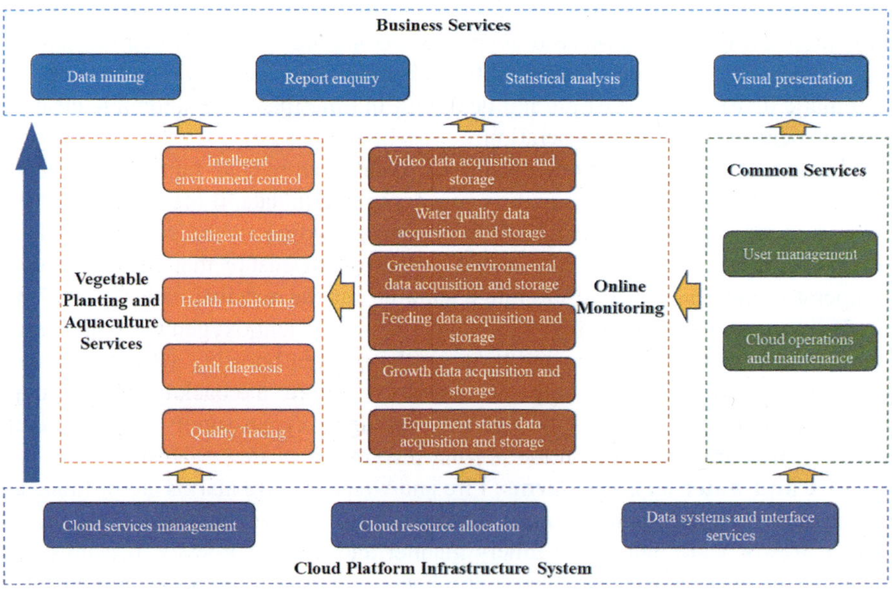

FIGURE 9.2

Functional service framework for cloud platform and control system.

various ways, for example, sensor and video data can be collected in real time and uploaded in batch, and its transmission interface adopts API, message queue or file transfer. Sensor and video data are uploaded using API, which is a software module that allows data exchange and communication interface between different applications for data input and output, and can realize instant data transmission and real-time processing. User data requests are accessed using message queuing, which is an asynchronous communication method used to decouple the sender and receiver of data by sending the data to the queue, which is then accessed and processed by the consumer in a sequential manner, to achieve efficient data transmission and processing. Large-scale data passthrough uses the File Transfer Protocol (FTP), which is a standard protocol used to transfer files over a computer network, and can use the FTP or other file transfer protocols to realize the interface for sending and receiving data by uploading and downloading files. The application program uses database connection interface to connect in order to realize the reading and writing operation of the data in the database, and the database connection interface has JDBC, ODBC, and so on, which can realize the efficient storage and retrieval of data. In the aquaponics system, sensor data, video streaming data, log files, user inputs, and so onare transmitted to the cloud platform through the data interface, where the data are cleaned, converted, and calculated to extract valuable information and insights, and the processed data are stored in the database or other storage media to facilitate rapid retrieval and querying of the data at a later stage and to provide data analysis and decision-making data support for data analysis and decision-making.

The cloud platform data system provides a variety of data-related functions and services that enable users to efficiently process and analyze data, including data storage, data processing, database management, data integration, data security and authority control, data visualization, and analysis, and real-time data processing.

Data storage: Provides a variety of data storage options, including cloud database, object storage, data lakes, and so on. The storage system can be scaled on-demand and provides highly reliable, secure, and persistent data storage.

Data processing: Supports batch and streaming processing of large-scale data. By using data processing engines or big data frameworks (such as Apache Hadoop and Apache Spark), data can be cleaned, transformed, aggregated, and calculated.

Database management: Provides a wide range of database options such as relational databases (e.g., MySQL, PostgreSQL, Oracle) and NoSQL databases (e.g., MongoDB, Cassandra, Redis) to meet different types and sizes of data storage and query requirements.

Data integration: Supporting the integration and consolidation of data from different sources, it can realize the functions of data extraction, transformation and loading (ETL), as well as data synchronization and replication to ensure the consistency and integrity of data.

Data security and privilege control: Provide data security and privilege control mechanisms, including authentication, access control, encryption, and auditing functions to protect the confidentiality and integrity of data.

Data visualization and analytics: Provides data visualization tools and analytics services that allow users to visualize and analyze data. This includes reports, charts, dashboards, and machine learning models.

Real-time data processing: The cloud platform supports real-time data stream processing, such as using streams like Apache Kafka or Amazon Kinesis.

9.3.2 Management system

The cloud platform management system is a software system used to manage and monitor the resources and services of a cloud platform, providing management and configuration functions for accounts and permissions, user task management, and application deployment.

9.3.2.1 *Account and privilege management*

Managing user accounts and access rights ensures that only authorized users can access and operate the cloud platform resources. Administrators can create and manage user accounts in the cloud platform management system, including setting user names and passwords for users and assigning them corresponding roles or rights. Administrators can create different roles and assign different combinations of permissions, for example, the administrator role may have full administrative privileges, while an ordinary user role can only partially operate on its own resources. Account authentication and authentication methods use account name passwords and keys to each other to ensure the legitimacy of the identity. ACLs are configured to restrict the access rights of specific users or user groups and prohibit certain users from modifying sensitive data, while administrators can view log records to understand users' access to and operation of cloud platform resources.

9.3.2.2 *User task management*

User task management is the main component of the aquaponics cloud platform, which includes seed culture management, equipment management, and system monitoring and alarm management, and each management module can call data or models to provide display functions, while users can create and set management parameters in the corresponding module.

9.3.2.2.1 Aquaculture management

The aquaculture management module covers five aspects: water cycle monitoring, feed management, fish management, growth monitoring management, and health monitoring and disease prevention. Water cycle monitoring is used to monitor the water quality parameters of the water body, including pH value, temperature, dissolved oxygen (Shao et al., 2025), ammonia nitrogen (Jiang et al., 2025), and so on, to monitor the water quality condition in real time, while the user can set the parameters of water quality requirements, water temperature range, and so on, and the system can regulate them according to the need to maintain a suitable water quality environment. Feed management mainly records and manages fish feed delivery, including feed type, dosage, and feeding frequency, to help control the nutritional intake and growth and development of fish, while users can set up feed delivery programs. Fish management is to record parameters such as aquaculture area (aquaculture area, source water area and tail water area), aquaculture workshop, aquaculture species, time, and so on. The aquaculture situation of a specific fish pond can be viewed in real time through video monitoring, and the user can set up key parameters such as fish and plant species, quantity and density (Wang et al., 2025), and the appropriate aquaculture system configurations and equipment combinations can be selected according to the actual situation. Growth monitoring and analysis is to collect and analyze fish growth data, such as weight growth, feed conversion rate, body color, swimming behavior, facility behavior, etc., to provide a basis for analyzing and understanding the fish growth trend and adjusting the culture strategy (Cui et al., 2024). Health monitoring and disease prevention monitor (Wang et al., 2025a) the health status of fish mainly through visual information, prevent disease outbreaks, and provide real-time alerts and warnings when necessary.

9.3.2.2.2 Planting management

The planting management module mainly includes four aspects: environment monitoring, fertilizer management, irrigation management, and growth data analysis. Environmental monitoring mainly provides the temperature, humidity, light and other parameters of the planting environment, while the user can set the optimal environmental parameters to help the system adjust the environmental conditions and maintain a suitable growing environment. Fertilizer management is to record the type, time and dosage of fertilizer application and provide visual data and reports to help managers apply fertilizer reasonably and improve plant growth efficiency. Irrigation management is to provide irrigation parameter information, which can be intelligently regulated according to the water demand of plants and the humidity of planting medium to ensure that plants get the right amount of water. Growth data analysis mainly records and provides plant growth data, such as growth rate, yield, leaf health status, nutritional status, and so on. By analyzing the plant growth status, it can adjust the management strategy in time.

9.3.2.2.3 Equipment management

The equipment management module is designed to monitor the status of individual equipment, and can set parameters for each equipment to realize manual remote management. The equipment management module includes six aspects: equipment monitoring and status, maintenance plan and record, fault diagnosis and repair, energy consumption management, remote operation and control, and data analysis and optimization. Equipment monitoring and status is mainly realized through sensors and video data to monitor the operation status and performance parameters of key equipment (e.g. irrigation system, water circulation equipment, environment control equipment, robots, etc.) in real time, and to provide information such as visual display and data report. Maintenance plan and record is used to establish and manage the regular maintenance plan of the equipment, record the maintenance activities and results, and provide information such as visual display and data report, which is mainly created and set by the management personnel. Fault diagnosis and repair is used to monitor the possible faults or abnormalities of the equipment, with visualization and data report display functions, providing real-time alarms, and supporting fault diagnosis and remote repair functions. Energy consumption management is used to monitor the energy consumption of the equipment, analyze and optimize the energy utilization efficiency, with visualization and data report display function. Remote operation and control are equipped with visualization and parameter setting functions to remotely access and control key equipment, robots, and so on, such as remotely adjusting the parameters of circulating water pumps or environmental control parameters to adapt to different planting and breeding needs. Data analysis and optimization is to collect and analyze the operating data of equipment or robots to optimize the equipment configuration and usage plan, with visualization and data report display function, and its optimization parameters are provided by optimization model or manual setting.

9.3.2.2.4 System monitoring and alarm management

System monitoring and alarm management is mainly for the possible existence of dangerous situations for parameter setting and alarm display, which provides monitoring parameter setting, alarm notification, and processing of three aspects, with parameter setting, visualization, and abnormal record report water yanking function. The monitoring parameters are set by the user, including type, range and threshold, such as water temperature, dissolved oxygen, and so on. When the monitoring data exceeds the set range or threshold, the system will trigger an alarm mechanism to remind the user of the existence of abnormal conditions. Once the system detects an anomaly and triggers an alarm, the system will send a notification to the management personnel and turn on the alarm system, and if necessary, control the local system to transfer to the disaster recovery operation state. When the anomaly is handled, the alarm message is contacted and the cause of the anomaly and the handling method are recorded to provide data support for later model analysis.

9.3.2.3 ***Application management***

Application program is the core component of aquaponics system to realize intelligence, it is the intelligent brain of aquaponics system, which provides analysis and decision-making ability for the system. The application programs of the aquaponics system cover a variety of models, including water quality prediction, sensor fault diagnosis, intelligent environment regulation, fish identification, fish behavior analysis, fish abnormal behavior diagnosis (Xu et al., 2024), fish health

(Zhao et al., 2022), and fish disease diagnosis, plant health status analysis, plant nutritional status analysis, intelligent equipment regulation, equipment health status analysis, energy consumption optimization, and so on. Each model is equipped with a separate visual display function, with automatic and manual recall. Each model has a separate visualization function, with automatic and manual call two ways.

The applications of aquaponics system are deployed in the cloud, and the deployment environment is divided into two types: test environment and production environment. Container technology (e.g., Docker) is used to package the applications and their dependencies into container images to realize cross-platform deployment and operation. Each application or updated application is first deployed to run in the test environment, and then deployed to the production environment after the test is passed and stabilized. Application deployment creates and configures the required application instances through the interface or command line tools provided by the cloud platform. The program needs to specify the application type, version, network settings, environment variables and other information. The application deployment protocols are HTTP/HTTPS, google remote procedure call (gRPC), and WebSocket. HTTP/HTTPS is used for secure communication and data transfer between the model service and the client, gRPC is used for efficient Remote Procedure Call (RPC), and cross-language communication, and WebSocket is used for bi-directional communication in real-time applications, such as real-time data updates and model prediction feedback. The cloud platform can be configured with monitoring and log management services to monitor the operational status and performance metrics of the application in real time and record critical log information, which helps to identify and resolve issues in a timely manner to ensure the stability and reliability of the application.

9.3.3 Operation and maintenance system

The O&M system is used to ensure the stable operation of the cloud platform, which includes hardware device management, monitoring and alerting, troubleshooting and recovery, security management, capacity planning, and system updates and upgrades.

9.3.3.1 Hardware equipment management

The hardware equipment of the hardware equipment management cloud platform includes servers, network equipment, storage equipment, and so on. By conducting regular inspections, maintenance and upgrades, it can ensure their normal operation and efficient utilization. By establishing an asset management system, the basic information, location, and configuration of the equipment are recorded. Reasonable planning of the life cycle of equipment, regular tracking and disposal of equipment obsolescence, ensuring safe decommissioning of equipment, and secure erasure of data.

9.3.3.2 Monitoring and alarms

Monitoring and alarming is an important link to ensure the stable operation of cloud platform. Taking server load, network traffic, storage utilization rate, application performance, and so on, as monitoring indexes, the monitoring tool or agent program is used to collect data of various indexes and store them into a special monitoring database or time series database, monitor the changes of the values of various indexes in real time through the monitoring system and set up corresponding alarm rules. When an indicator reaches or exceeds the set threshold, the system will trigger an alarm and notify the manager by instant messaging or cell phone push. At the same time, the monitoring data analysis and trend prediction model is configured to understand the performance status and potential problems of the system, predict the trend of resource demand and load changes in order to make corresponding optimization and planning, and discover potential optimization space by collecting historical data, conducting performance analysis and comparison, and adjusting the platform configuration and resource allocation in a timely manner, so as to enhance the stability and performance of the system.

9.3.3.3 Fault management and recovery

Troubleshooting and recovery are key aspects of ensuring the availability of cloud services. Through the establishment of an effective monitoring system and alarm mechanism, the status of each component and resource of the cloud platform is monitored in real time, and an alarm is triggered and the administrator is notified in a timely manner in case of abnormality or failure. Based on the alarm information and log data, the administrator quickly locates the specific component or resource that has failed, and uses log analysis, network packet capture and other troubleshooting tools and techniques to help determine the cause of the failure and repair it. After completing fault repair, service recovery and validation work is then performed to ensure that the repaired service can operate normally and to verify that users can access and use the cloud service normally (Palm et al., 2018). After fault recovery, information such as the cause of the fault, the repair process, and the response time need to be recorded so that potential system problems can be identified through fault analysis and appropriate preventive measures can be taken to improve the stability and reliability of the cloud platform.

9.3.3.4 Cloud security management

Cloud platform security management is an important task to ensure the security of data and systems in a cloud environment through authentication, access rights management, and strong password policies to ensure that only authorized users can gain access. Encrypt sensitive data, including encryption during data transmission (e.g., HTTPS) and encryption during data storage (e.g., database encryption, file storage encryption) to prevent data from being accessed or leaked by unauthorized personnel. Establish a detailed logging and auditing mechanism to monitor the operational behavior and events of the cloud platform, detect abnormal activities and security threats in a timely manner, and use technologies such as security information and event management systems to conduct real-time monitoring and analysis to provide a rapid response. Employ security equipment such as firewalls, intrusion detection, and prevention systems (IDS/IPS) to monitor and block malicious network traffic and attacks, conduct regular vulnerability scanning and security assessments, and identify potential vulnerabilities and take remedial measures at an early stage. Establish a disaster recovery and backup mechanism to ensure data disaster tolerance and recovery capability, and improve system availability and disaster resistance through backup data and fault-tolerance technology.

9.3.3.5 Capacity planning

As a result of model and business expansion, the cloud platform needs to estimate resource usage based on demand to ensure that the cloud environment can meet the resource requirements for computing, storage, and network needed for operation. First, it needs to accurately understand the business requirements and expected growth trends, including the number of users, data volume, access frequency, and so on, and reasonably estimate the computing resources, storage capacity, and network bandwidth, based on the business requirements. Second, the cloud platform should adopt elastic scaling mechanism, automatic load balancing and other technical means to automatically expand capacity during peak hours or business growth, while releasing resources during low peak periods to improve resource utilization and cost-effectiveness. Then, through stress testing and performance monitoring of key applications, bottlenecks are identified and corresponding capacity planning and optimization measures are carried out. Finally, future development needs should be considered and a certain amount of expansion space should be reserved so that resources can be expanded in a timely manner when business grows, avoiding performance problems and degradation of user experience caused by insufficient capacity.

9.3.3.6 System update and upgrade

System update and upgrade is an important operation to improve system performance, perfect functions and fix vulnerabilities. Timely obtain software and system patches provided by vendors and conduct adequate testing and verification work, including functionality testing, performance testing, compatibility testing, etc., to ensure the stability and compatibility of the new version before system updates and upgrades. At the same time, make a full backup of the system before updating and set up a rollback policy in case there are problems in the updating process or it does not meet expectations.

9.4 Typical case: Chongqing aquaponics artificial intelligence factory

The Aquaponics Artificial Intelligence Factory of Chongqing Academy of Agricultural Sciences is China's first aquaponics factory production technology and equipment innovation and research and development platform, including intelligent planting workshop of 4000 square meters, processing pilot plant of 1200 square meters, fish breeding workshop of 1850 square meters, and water and fertilizer control workshop of 300 square meters (Chongqing Daily, 2022). Fish and vegetable production area and scale mainly includes vegetable cultivation area of 2000 square meters, vegetable nursery area of 1000 square meters, aquaculture water body of 800 cubic meters, with an annual output of 100 tons of green vegetables and 160 tons of large mouth black bass. The aquaponics artificial intelligence factory adopts a double-layer structure design, with one layer for high-density circulating water aquaculture and the second layer for vegetable cultivation, and the whole factory includes a circulating water fish farming system, a precision baiter system, a fish manure concentration and biochemical treatment system, an unmanned vegetable nursery system, a greenhouse intelligent logistics system, and a spectrophotometric three-dimensional cultivation system. It is equipped with water sensors, video and audio sensors, and a video and audio system. It is equipped with automated equipment such as water quality sensors, video monitoring, automatic seeders, intelligent working robots, and greenhouse intelligent transportation robots.

The Aquaponics Artificial Intelligence Factory enables intelligent and unmanned operation of production processes such as vegetable seeding, nursery, transplanting and planting, logistics, irrigation, environmental control, harvesting, cutting, packaging, fish rehydration, baiting and feeding, feeding, temperature control, monitoring, disinfection, recycling, sewage and feces collection, and treatment.

The high-density circulating water aquaculture and greenhouse environment control system adopts the "cloud-network-end" mode to realize intelligent and precise regulation and control. The data center of the cloud platform analyzes and makes decisions by collecting data from intelligent equipment and production scenarios, controls the operation of the circulating water system, biochemical system, greenhouse environment control equipment and operating robots, realizes the regulation and control of the water quality and environmental parameters of fish farming, and completes the intelligent operation of water replenishment, feeding, oxygenation, temperature control, and manure treatment. Intelligent feeding is realized by giving decision-making instructions from the monitoring platform and controlling the crawler feeding robot to feed bait automatically at a fixed point and in a fixed quantity; fishing and grading is realized by controlling the automatic fish-sorting equipment from the monitoring platform.

In the vegetable planting area, the unmanned operation of "one seed to one vegetable" was realized for the first time. Seedlings were sown into seedling trays through the fully automated seeding assembly line and potting and transplanting and planting equipment, and transported to the unmanned tidal seedling nursery system by the intelligent operation robot. Next, the intelligent operation robot and autonomous transportation robot transported the seedlings to the spectral stereoscopic cultivation frame for further planting. Finally, with the help of an intelligent harvesting system for leafy vegetables, the ripe vegetables are picked by an intelligent harvesting robot and transported by an intelligent transportation robot to the harvesting workshop for cutting and packaging. The whole process of vegetable cultivation is controlled by the intelligent monitoring platform of the aquaponics AI factory, and the operating system receives instructions to execute the corresponding operation process.

9.5 Conclusion

Aquaponics smart factories require hardware and software facilities to cooperate with each other, and the core of which is the reasonable deployment of intelligent equipment and decision-making models. Most of the aquaponics intelligent factories that have been built so far adopt the structure of "cloud-network-end," which requires high computation on the cloud platform, especially in the case of a large amount of data, such computation will be immeasurable, which will lead to delays in the control instructions, thus affecting the operation of the system. In addition, the output rate and stability of the network will also affect the final control effect, so in the future, we can increase the application of edge computing, especially when it comes to the local control of equipment, operation of robots, etc., the use of "cloud-network-edge-end" mode can be more effective to improve operational efficiency (Karimanzira and Jacobi, 2022).

Although many predictive analysis models have been studied for fish sensors, devices, and behavioral analysis, it is undeniable that these models have some special requirements for input data, and huge changes in the environment can lead to model failure, while high-precision models need to consume more deployment resources. With the application of large-scale models, constructing a multimodal-based large-scale model decision-making strategy will help to realize a more accurate regulation of aquaponics intelligent system.

Data transmission with high throughput and high transmission rate is still the best communication method pursued by smart factories, and 5G transmission has met most of the requirements, but the application of aquaponics system based on 5G data transmission still needs to be further promoted.

References

Chongqing Daily. (2022, September 14). Aquaponics AI factory, seedling factory, unmanned orchard, Chongqing's smart agriculture has been "revolutionized". https://www.cq.gov.cn/ywdt/jrcq/202209/t20220914_11106569.html.

Cui, M., Liu, X., Liu, H., Du, Z., Chen, T., Lian, G., ... Wang, W., 2024. Multimodal fish feeding intensity assessment in aquaculture. IEEE Transactions on Automation Science and Engineering.

Goddek, S., Delaide, B., Mankasingh, U., Ragnarsdottir, K.V., Jijakli, M.H., Thorarinsdottir, R., 2015. Aquaponics: an industrial application of closed loop systems. Systems 3 (3), 33–55.

Jiang, L., Jia, M., Guo, B., Li, D., Chen, T., 2025. Mathematical modelling of ammonia nitrogen dynamics in a recirculating aquaculture system. Aquacultural Engineering 102594.

Karimanzira, D., Jacobi, M., 2021. Electronic management of a recirculating aquaculture system. Aquacultural Engineering 92, 102143.

Karimanzira, D., Jacobi, M., 2022. Digital twin-based control and monitoring of an aquaponics system. Computers and Electronics in Agriculture 197, 106967.

Kyaw, T.Y., Ng, A.K., 2017. Smart aquaponics system for urban farming in Singapore. Proceedings of the IEEE International Conference on Industrial Engineering and Engineering Management (IEEM), 1141-1145.

Li, Z., Wang, J., Higgs, R., Zhou, J., Tan, C.J., 2020. Design of an intelligent aquaponics monitoring system based on IoT and deep learning. Journal of Sensors 2020, 1–15.

Mahon, N., Raman, S., Fraile-Navarro, D., 2022. The role of artificial intelligence in sustainable agriculture: a systematic literature review. Journal of Cleaner Production 352, 131537.

Mobley, R.K., 2002. An Introduction to Predictive Maintenance. Butterworth-Heinemann.

Palm, H.W., Knaus, U., Appelbaum, S., Goddek, S., Strauch, S.M., Vermeulen, T., et al., 2018. Towards commercial aquaponics: a review of systems, designs, scales and components. Fisheries and Aquatic Sciences 21 (1), 1–16.

Somerville, C., Cohen, M., Pantanella, E., Stankus, A., Lovatelli, A., 2014. Small-scale aquaponic food production: integrated fish and plant farming. FAO Fisheries and Aquaculture Technical Paper No 589.

Shao, K., Li, D., Tang, H., Zhang, Y., Xu, B., Bhatti, U.A., 2025. Improving multi-step dissolved oxygen prediction in aquaculture using adaptive temporal convolution and optimized transformer. Computers and Electronics in Agriculture 235, 110329.

Wei, Y., Li, W., An, D., Li, D., Jiao, Y., Wei, Q., 2019. Equipment and intelligent control system in aquaponics: a review. IEEE Access 7, 169306–169326.

Wang, J., Yin, Y., Yang, J., Zhu, F., Li, D., Wang, Y., 2025a. Real-time rapid visual fish disease detection system based on tiny machine learning. Aquaculture International 33 (6), 1–22.

Wang, G., Yu, J., Xu, W., Muhammad, A., Li, D., 2025b. Automated fish counting system based on instance segmentation in aquaculture. Expert Systems with Applications 259, 125318.

Xu, W., Liu, C., Wang, G., Zhao, Y., Yu, J., Muhammad, A., et al., 2024. Behavioral response of fish under ammonia nitrogen stress based on machine vision. Engineering applications of artificial intelligence 128, 107442.

Yanes, A.R., Martinez, P., Ahmad, R., 2020. Real-time monitoring and control of an aquaponics system using IoT and PLC. Journal of King Saud University–Computer and Information Sciences 32 (9), 1030–1039.

Zhao, S., Zhang, S., Lu, J., Wang, H., Feng, Y., Shi, C., ... Zhao, R., 2022. A lightweight dead fish detection method based on deformable convolution and YOLOV4. Computers and Electronics in Agriculture 198, 107098.

CHAPTER

Sustainability evaluation of smart aquaponics farming 10

Daoliang Li[1,2], Youbang Guan[3] and Lihua Song[4]

[1]*National Innovation Center for Digital Fishery, China Agricultural University, Beijing, P.R. China,*
[2]*College of Information and Electrical Engineering, China Agricultural University, Beijing, P.R. China,*
[3]*Centre for Environment and Sustainability, School of Sustainability, Civil and Environmental Engineering, University of Surrey, Guildford, United Kingdom,*
[4]*School of Chemistry & Chemical Engineering, University of Surrey, Guildford, United Kingdom*

Aquaponics is an advanced agricultural method that blends aquaculture with hydroponics, creating a system where water and nutrients are recycled efficiently to support both fish and plant growth. This innovative approach has attracted attention because it not only conserves resources but also enhances efficiency and provides significant environmental benefits. As the global population grows and environmental challenges intensify, aquaponics holds great promise for boosting food production while minimizing ecological impact, positioning itself as an important contributor to sustainable development.

The sustainability of aquaponics can be analyzed through four key lenses: economic feasibility, environmental conservation, carbon neutrality, and social impact. A comprehensive review of these factors reveals the system's advantages and offers valuable insights for policymakers and stakeholders interested in encouraging broader implementation.

The goal of this chapter is to thoroughly evaluate the sustainability of aquaponics by examining its contributions at economic, environmental, carbon-reducing, and social dimensions. Through this analysis, we aim to highlight how aquaponics can be a viable solution to some of the pressing environmental and societal challenges of today. This chapter is structured as follows:

- Economic evaluation: Focuses on the costs, potential revenues, and return on investment associated with aquaponics systems.
- Ecological evaluation: Looks into how aquaponics reduces pollution, promotes waste recycling, and optimizes water use.
- Net-zero evaluation: Evaluates the system's capacity to reduce carbon emissions, nitrogen oxides, and improve nutrient management.
- Social benefit evaluation: Examines the benefits of land-use efficiency, improved productivity, reduced labor needs, and positive community effects.

Smart Aquaponic Farming. DOI: https://doi.org/10.1016/B978-0-443-28945-3.00006-2

By exploring these dimensions, this chapter aims to demonstrate the broad benefits and sustainability of aquaponics, providing strong support for its broader application.

10.1 Economic evaluation

The initial setup of an aquaponics system involves costs related to essential equipment such as fish tanks, grow beds, water pumps, and filtration systems, with expenses varying based on the size and complexity of the system (Rakocy, 2012). In addition, incorporating advanced features like automation and sensors can increase the overall cost but offers the benefit of more efficient system management and higher production yields (Yanes et al., 2020). Key biological inputs for the system include fish stock and plant seeds, both of which are critical to ensuring the system's productivity. Initial cost is directly related to the type of selected fish and plants. High-quality fingerlings and seeds can increase the system production efficiency and profitability (Bosma et al., 2017).

Aquaponics systems have a clear advantage in efficient resource utilization. Water and nutrients are recycled within the system, significantly reducing the consumption of water resources and fertilizers. Aquaponics is more profitable in the long term than conventional agriculture, saving up to 85% of water wasted by conventional farming techniques in Egypt (El-Essawy et al., 2019). Aquaponic systems utilize natural biological nitrogen cycles, minimizing nonrenewable resource use, and providing economic benefits over time. Sustainable opportunities include biological nitrogen production rates of 80–90 g/m^3 per day nitrate nitrogen from trickling biofilters and plant uptake of aquaculture wastewater. This system results in improved water and nutrient use efficiency and conservation (Tyson et al., 2011).

By recycling nutrients, aquaponics systems reduce agricultural wastewater discharge, thus lowering environmental management costs. The system generally requires no added chemicals, reducing the risk of water pollution and the associated environmental management expenses. Fish farming provides a significant source of income in aquaponics systems. Depending on market demand and the species of fish, revenue from fish sales can be substantial. Hydroponically grown vegetables typically have fast growth rates and high quality, which allows them to fetch higher prices in the market. For instance, small-scale aquaponic systems produced 62 kg of tilapia and 352 kg of vegetables and fruits, with an average net agricultural added value of €151.3. Choosing fish species in high demand can maximize profits (Lobillo-Eguíbar et al., 2020).

Aquaculture systems are 29% more efficient at raising fish and five times more efficient at producing pumpkin than conventional systems, with better water quality and higher fish survival rates (Oladimeji et al., 2020). Cost-effective aquaponics systems can enhance food security and provide income for farming households in coastal Bangladesh, with a benefit-cost ratio of 2.2 (Sunny et al., 2019). Decoupled aquaponic systems using fish water for lettuce production can save 62.8% of inorganic fertilizers and reduce greenhouse gas emissions by 72% compared to conventional hydroponic systems (Monsees et al., 2019). By-products from fish and plants, such as fish fertilizer and plant residues, can also be processed or sold to generate additional income, and these by-products can be used in gardening or sold as bio-fertilizers.

Fish feed is the primary operating cost in an aquaponics system, accounting for more than 50% of the total operating expenses (Vassiliou et al., 2015). Energy consumption includes the electricity costs for water pumps, lighting, and temperature control systems, with the electricity cost depending on lighting requirements and usage frequency. Choosing energy-efficient equipment can further reduce energy consumption. For example, fish feed is a primary operating cost in small-scale raft aquaponic systems, with an average cost of $18 per kg of fish. The average energy use was 19,526 kWh for propane and electricity per year at a cost of $2055 in Baltimore, Maryland, United States (Love et al., 2015). Daily maintenance includes cleaning fish tanks, checking pipes, replacing water pumps, and pruning plants, which mainly involves low labor costs.

Technological advancements and system optimization play a crucial role in boosting the economic efficiency of aquaponics systems. For example, using high-efficiency water pumps and energy-saving lighting can significantly reduce energy consumption (Fang et al., 2017). In addition, optimizing bio-filters enhances nutrient utilization, cutting down the need for external fertilizers (Zou et al., 2016). The integration of automated control systems can further increase production efficiency while lowering labor costs (Karimanzira and Rauschenbach, 2019).

As the scale of the system expands, both initial equipment and operating costs can be reduced. Large-scale systems benefit from economies of scale, allowing for more cost-effective equipment procurement and operational efficiencies, which in turn improves overall economic viability (Greenfeld et al., 2019). By expanding the system, unit costs decrease, leading to increased profitability. However, the payback period for small-scale home-based aquaponic systems may take 5 years or more, depending on estimated future costs, benefits, and discount rates (Tokunaga et al., 2015). Effective management and strategic market positioning can help shorten this payback period.

The long-term financial sustainability of aquaponics systems relies on consistent profitability. With the dual production of fish and vegetables and efficient use of resources, aquaponics systems offer a steady revenue stream. Sustained market demand and a well-thought-out pricing strategies are key to maintaining profitability (Tokunaga et al., 2015). In addition, adapting to changing market conditions, such as shifts in consumer preferences and fluctuations in demand, is critical for long-term success. Government policy support, including financial subsidies and tax incentives for sustainable agriculture, also plays an important role in securing the long-term viability of aquaponics systems (Lobillo-Eguíbar et al., 2020). For instance, applying for funding to support green agricultural projects can help reduce initial investment risks. Understanding and leveraging available government resources and support mechanisms can greatly improve the financial sustainability of the system.

10.2 Ecological evaluation

The ecological evaluation of aquaponics systems centers around several critical metrics, including the reduction of ammonia nitrogen pollution, effective waste recycling, efficient use of water resources, and the maintenance of biodiversity and ecological balance. Each of these factors plays a significant role in highlighting the environmental benefits and sustainability of aquaponics systems, offering a comprehensive view of how they contribute to reducing environmental impact while promoting resource conservation and ecosystem health.

10.2.1 Reduction of ammonia nitrogen pollution

Aquaponics systems effectively reduce ammonia nitrogen pollution through natural biological processes (Rakocy, 2012). Beneficial bacteria present in the biofilter convert ammonia, which is produced from fish waste, into nitrites and then into nitrates. As noted by Tyson et al. (2011), these nitrates are absorbed by plants as nutrients, thus significantly lowering harmful ammonia concentrations in the water.

To monitor this reduction, regular water samples are taken from various parts of the system, such as fish tanks, biofilters, and plant beds. These samples are analyzed using techniques like spectrophotometry or ion-selective electrodes, allowing for precise measurement of ammonia nitrogen levels. The collected data is then compared to baseline readings from traditional aquaculture systems to assess the reduction in ammonia nitrogen pollution.

Key metrics in this analysis include the concentration of ammonia (mg/L) before and after biofilter treatment (Endut et al., 2011). For instance, an urban aquaponics project in Melbourne, Australia, demonstrated a 60% reduction in ammonia nitrogen levels compared to traditional aquaculture systems. The biofilter played an essential role in this process by converting ammonia into nitrates, which were absorbed by the plants (Goddek et al., 2015). This case illustrates the effectiveness of aquaponics in mitigating ammonia nitrogen pollution, thereby contributing to a healthier aquatic environment.

10.2.2 Waste recycling

Waste recycling is a major ecological benefit of aquaponics systems. Fish waste, which would normally contribute to environmental pollution, is converted into valuable nutrients that support plant growth (Rakocy, 2012). This recycling process is measured by monitoring the input of fish feed and the output of fish waste. Solid waste collected from the fish tanks is analyzed for its nutrient content, particularly nitrogen, phosphorus, and potassium. The efficiency of converting waste into nutrients for plants is assessed by comparing the nutrient input from fish feed with the nutrient uptake by plants.

Key performance metrics include the percentage of waste recycled and the nutrient levels found in plant beds (Somerville et al., 2014). For example, the urban aquaponics project in Melbourne achieved a 95% waste recycling rate, substantially reducing the amount of organic waste produced (Love et al., 2015). Similarly, a greenhouse aquaponics system in Vancouver, Canada, reported a 98% waste recycling efficiency, resulting in minimal waste production and improved plant growth and productivity (Love et al., 2015). These examples showcase the potential of aquaponics systems to efficiently recycle waste and optimize nutrient use (Graber and Junge, 2009).

In Sweden, a sustainable aquaponics system near Stockholm used advanced biofiltration techniques to significantly reduce ammonia nitrogen levels. This system, powered by renewable energy sources such as wind and solar, reduced ammonia nitrogen concentrations by 70% (Goddek et al., 2015). This case highlights the potential of combining renewable energy with aquaponics to further enhance environmental sustainability.

10.2.3 Biodiversity and ecological balance

One of the key benefits of aquaponics systems is their exceptional efficiency in water resource utilization. Compared to conventional soil-based agriculture, aquaponics uses around 90% less water, thanks to its high water-recycling rates (Somerville et al., 2014). To assess water resource utilization efficiency, flow meters are installed at multiple points within the system to monitor water use continuously. The water footprint for each kilogram of produce, whether fish or plants, is calculated by dividing the total water used by the total weight of produce harvested. This efficiency is then compared with that of traditional soil-based agriculture (Endut et al., 2011). Key metrics include total water usage (liters) and the water recycling rate (Love et al., 2015).

For example, the Melbourne urban aquaponics project achieved a 95% waste recycling rate, which significantly reduced organic waste output. Similarly, a greenhouse aquaponics system in Vancouver, Canada, focusing on sustainability, reported a 98% waste recycling efficiency, minimizing waste and improving plant productivity. A low-emission aquaponics project in Auckland, New Zealand, optimized waste recycling processes to achieve a 98% waste recycling rate, effectively converting fish waste into nutrients for plants and substantially reducing organic waste output. These examples highlight the efficiency of aquaponics systems in promoting waste recycling and nutrient utilization (Somerville et al., 2014).

Aquaponics systems also promote ecological balance and biodiversity by cultivating a diverse array of species. Regular inventories are conducted to monitor the diversity and health of both fish and plant species (Graber and Junge, 2009). Indicators such as plant health and fish growth rates provide insights into the system's ability to maintain a balanced ecosystem (Timmons & Ebeling, 2013). Biodiversity indices, like the Shannon–Weaver index, are used to quantify species diversity (Somerville et al., 2014). Ecological monitoring involves periodic assessments of plant vitality and fish health, tracking factors such as growth rates and disease prevalence (Endut et al., 2011).

In New Zealand, the low-emission aquaponics project in Auckland maintained a high level of biodiversity by incorporating a variety of fish and plant species. Regular ecological assessments showed a healthy and balanced ecosystem, with strong growth rates and minimal occurrences of disease. This project demonstrates the ability of aquaponics systems to support and enhance biodiversity (Goddek et al., 2015).

10.3 Net-zero assessment

Aquaponics systems offer the potential to greatly reduce carbon emissions when compared to traditional farming methods (Timmons and Ebeling, 2013). This reduction is primarily driven by the efficient use of resources, such as water and nutrients, and the incorporation of renewable energy sources. By minimizing waste, recycling nutrients, and relying on sustainable energy options, aquaponics systems can help mitigate the environmental impact of agricultural activities, contributing to a more sustainable and lower-carbon food production model.

10.3.1 Carbon emissions and carbon footprint

To measure carbon emissions in aquaponics systems, carbon footprint analysis is conducted by calculating the total greenhouse gas emissions (CO_2 equivalent) produced by the system. This includes emissions from energy consumption, transportation, and inputs such as fish feed and plant nutrients. Tools like life cycle assessment software can provide a comprehensive evaluation of carbon emissions throughout the system's entire lifecycle. Key metrics used in this analysis include total CO_2 emissions (kg CO_2e) and emissions per unit of produce (kg CO_2e/kg of produce) (Somerville et al., 2014).

For example, a sustainable aquaponics system in Sweden integrated renewable energy sources like wind and solar power, significantly reducing its dependence on fossil fuels. The carbon footprint analysis for this system revealed a 60% reduction in carbon emissions compared to conventional farming systems, demonstrating the effectiveness of renewable energy in achieving net-zero carbon emissions

(Timmons and Ebeling, 2013). This example illustrates how integrating sustainable energy solutions can drastically lower the carbon footprint of aquaponics systems.

10.3.2 Nitrogen oxides emissions

Nitrogen oxides (NOx) are significant pollutants responsible for environmental issues such as acid rain and photochemical smog. Aquaponics systems can effectively reduce NOx emissions through optimized nutrient management and waste recycling processes. Gas analyzers are used to measure NOx concentrations in both the effluent and air within the system. These measurements are taken at different stages, including fish tanks, biofilters, and plant beds. Key metrics include NOx concentrations (ppm) and total NOx emissions (kg/year).

By comparing aquaponics systems with conventional farming, reductions in NOx emissions can be accurately quantified. For instance, a low-emission aquaponics project in Auckland, New Zealand, implemented advanced nutrient management techniques, leading to a 70% reduction in NOx emissions compared to traditional agricultural methods (Love et al., 2014). The system's efficient conversion of nitrogen compounds was a critical factor in achieving this reduction (Goddek et al., 2015).

Aquaponics systems are designed to minimize nutrient loss by recycling nutrients within a closed-loop system. This not only enhances plant growth but also prevents the contamination of surrounding ecosystems, making the system more environmentally sustainable (Graber and Junge, 2009).

10.3.3 Nutrient loss

N nutrient loss in aquaponics systems is evaluated by analyzing the nutrient content in the system's effluent and comparing it with the initial nutrient input. Regular sampling of water and plant tissues is carried out to measure essential nutrients, such as nitrogen, phosphorus, and potassium. Tools like spectrophotometry and atomic absorption spectroscopy are used to conduct precise nutrient analysis. Key metrics for this assessment include nutrient retention rates (%) and nutrient loss (kg/year) (Somerville et al., 2014).

For instance, a greenhouse aquaponics system in Vancouver, Canada, achieved nutrient retention rates of over 95%, illustrating the system's high efficiency in recycling nutrients (Love et al., 2014). This minimal nutrient loss helped avoid potential contamination of nearby water bodies, thereby contributing to environmental sustainability.

In Sweden, a sustainable aquaponics system located near Stockholm utilized advanced biofiltration combined with renewable energy sources, resulting in significant reductions in both carbon and nitrogen oxide emissions (Goddek et al., 2015). The system's carbon footprint was reduced by 60%, while NOx emissions were lowered by 70% compared to conventional farming methods

(Graber and Junge, 2009). Effective nutrient management also ensured minimal nutrient loss, further enhancing the system's environmental sustainability (Timmons and Ebeling, 2013).

Similarly, a low-emission aquaponics project in Auckland, New Zealand, optimized nutrient management and incorporated renewable energy sources, leading to impressive reductions in environmental pollutants. The project successfully reduced carbon emissions by 50% and NOx emissions by 70%, while maintaining nutrient retention rates of over 95%. This demonstrates the system's effectiveness in minimizing environmental impacts and supporting sustainable agricultural practices (Graber and Junge, 2009).

These case studies highlight how aquaponics systems can contribute to achieving net-zero emissions and play a crucial role in advancing sustainable agriculture.

10.4 Social benefit evaluation

Aquaponics systems can yield food efficiently in limited spaces. In contrast to conventional farming, which typically requires large areas of land, aquaponics can be conducted on small plots or land unsuitable for farming. This makes the system especially suited to urban areas or regions with limited land resources (Wei et al., 2019). Compared to traditional farming, aquaponics significantly increases yield per unit area. By combining aquaculture with hydroponics, the system maximizes resource recycling. The proposed aquaponic system, which produces 60,000 ornamental fish and 108,000 lettuce heads per year, would impact the environment at an annual cost of 10,700 EUR, about half the environmental cost of producing these items separately (Greenfeld et al., 2021). This high yield can effectively address food shortages and meet urban food demands.

Aquaponics systems use many automated and semi-automated technologies, such as automatic feeding, intelligent water quality monitoring, and automatic irrigation. These technologies significantly decrease the physical labor required to run aquaponics systems. For example, automatic feeding systems can distribute feed at set intervals without human intervention, allowing operators to focus more on optimizing and managing the system efficiently (Pu'Ad et al., 2020). The use of modern technology and efficient management greatly enhances labor productivity in aquaponics systems. In comparison, traditional farming requires more labor to achieve similar production levels. This efficient labor use not only reduces operating costs but also increases overall production efficiency.

Aquaponics systems also serve as valuable educational tools for schools and communities, offering opportunities for science education, environmental education, and vocational training. By participating in the system's operation, students and community members can learn about agricultural production, ecological balance, and sustainable development. This educational approach stimulates interest in learning while developing practical skills and raising environmental awareness. Furthermore, aquaponics systems can grow various vegetables and fish, contributing to the preservation and promotion of local and traditional food cultures. By cultivating local vegetable varieties and raising indigenous fish species, aquaponics enriches the dietary culture and food options of the community. In addition, aquaponics can showcase regional agricultural techniques and practices, promoting cultural diversity and social inclusion (Wirza and Nazir, 2021).

Aquaponics systems can also enhance local economic growth. The sale of fish and vegetables provides communities with additional income, which helps improve living standards and generates tax revenue for local governments. This, in turn, supports public services and infrastructure development. For instance, introducing aquaponics systems in urban areas can create new markets, attract consumers and tourists, and stimulate local economic development.

Aquaponics systems represent a significant advancement in modern agricultural technology, contributing to the modernization of regional agriculture. This technology can be integrated with other modern farming practices, such as precision farming and smart agriculture, to enhance efficiency and sustainability in agricultural production. Promoting aquaponics technology can facilitate the transition from traditional farming methods to modern, efficient agricultural models, ultimately boosting agricultural productivity in the region.

AI disclosure

During the preparation of this work the author(s) used chatgbt in order to Language polishing. After using this tool/service, the author(s) reviewed and edited the content as needed and take(s) full responsibility for the content of the publication.

References

Bosma, R.H., Lacambra, L., Landstra, Y., Perini, C., Poulie, J., Schwaner, M.J., et al., 2017. The financial feasibility of producing fish and vegetables through aquaponics. Aquacultural Engineering 78, 146–154. Available at https://doi.org/10.1016/j.aquaeng.2017.07.002.

El-Essawy, H., Nasr, P., Sewilam, H., 2019. Aquaponics: a sustainable alternative to conventional agriculture in Egypt – a pilot scale investigation. Environmental Science and Pollution Research 26 (16), 15872–15883. Available at https://doi.org/10.1007/s11356-019-04970-0.

Enduta, A., Jusoh, A., Ali, N., Wan Nik, W.B., 2011. Nutrient removal from aquaculture wastewater by vegetable production in aquaponics recirculation system. Desalination and Water Treatment 32 (1-3), 422–430. Available at https://doi.org/10.5004/dwt.2011.2761.

Fang, Y., Hu, Z., Zou, Y., Fan, J., Wang, Q., Zhu, Z., 2017. Increasing economic and environmental benefits of media-based aquaponics through optimizing aeration pattern.

Journal of Cleaner Production 162, 1111–1117. Available at https://doi.org/10.1016/j.jclepro.2017.06.158.

Goddek, S., Delaide, B., Mankasingh, U., Ragnarsdottir, K., 2015. Challenges of Sustainable and Commercial Aquaponics. Sustainability 7 (4), 4199–4224. Available at https://doi.org/10.3390/su7044199.

Graber, A., Junge, R., 2009. Aquaponic Systems: Nutrient recycling from fish wastewater by vegetable production. Desalination 246 (1-3), 147–156. Available at https://doi.org/10.1016/j.desal.2008.03.048.

Greenfeld, A., Becker, N., McIlwain, J., Fotedar, R., Bornman, J.F., 2019. Economically viable aquaponics? Identifying the gap between potential and current uncertainties. Reviews in Aquaculture 11 (3), 848–862. Available at https://doi.org/10.1111/raq.12269.

Greenfeld, A., Becker, N., Bornman, J.F., Spatari, S., Angel, D.L., 2021. Monetizing environmental impact of integrated aquaponic farming compared to separate systems. Science of the Total Environment 792. Available at https://doi.org/10.1016/j.scitotenv.2021.148459.

Karimanzira, D., Rauschenbach, T., 2019. Enhancing aquaponics management with IoT-based predictive analytics for efficient information utilization. Information Processing in Agriculture 6 (3), 375–385. Available at https://doi.org/10.1016/j.inpa.2018.12.003.

Lobillo-Eguíbar, J., Fernández-Cabanás, V.M., Bermejo, L.A., Pérez-Urrestarazu, L., 2020. Economic sustainability of small-scale aquaponic systems for food self-production. Agronomy 10 (10). Available at https://doi.org/10.3390/agronomy10101468.

Love, David C., Fry, Jillian P., Genello, Laura, Hill, Elizabeth S., Frederick, J. Adam, Li, Ximin, Semmens, Ken, 2014. An International Survey of Aquaponics Practitioners. PLoS ONE 9 (7), e102662. Available at https://doi.org/10.1371/journal.pone.0102662.

Love, D.C., Uhl, M.S., Genello, L., 2015. Energy and water use of a small-scale raft aquaponics system in Baltimore, Maryland, United States. Aquacultural Engineering 68, 19–27. Available at https://doi.org/10.1016/j.aquaeng.2015.07.003.

Monsees, H., Suhl, J., Paul, M., Kloas, W., Dannehl, D., Würtz, S., 2019. Lettuce (*Lactuca sativa*, variety Salanova) production in decoupled aquaponic systems: same yield and similar quality as in conventional hydroponic systems but drastically reduced greenhouse gas emissions by saving inorganic fertilizer. PLoS One 14 (6). Available at https://doi.org/10.1371/journal.pone.0218368.

Oladimeji, S.A., Okomoda, V.T., Olufeagba, S.O., Solomon, S.G., Abol-Munafi, A.B., Alabi, K.I., et al., 2020. Aquaponics production of catfish and pumpkin: comparison with conventional production systems. Food Science and Nutrition 8 (5), 2307–2315. Available at https://doi.org/10.1002/fsn3.1512, onlinelibrary.wiley.com/journal/10.1002/%28ISSN%292048-7177.

Pu'Ad, M.F.M., Sidek, K.A., Mel, M., 2020. Automated aquaponics maintenance system. J. Phys. Conf. Ser. 1502.

Rakocy, J.E., 2012. Aquaponics-integrating fish and plant culture. Aquaculture Production Systems. Wiley-Blackwell, Virgin Islands (U.S.), pp. 344–386.

Somerville, Christopher, et al., 2014. Small-Scale Aquaponic Food Production- Integrated Fish and Plant Farming. FAO Fisheries and Aquaculture, pp. 262.

Sunny, A.R., Islam, M.M., Rahman, M., Miah, M.Y., Mostafiz, M., Islam, N., et al., 2019. Cost effective aquaponics for food security and income of farming households in coastal Bangladesh. Egyptian Journal of Aquatic Research 45 (1), 89–97. Available at https://doi.org/10.1016/j.ejar.2019.01.003.

Timmons, M.B., Ebeling, J.M., 2013. Recirculating Aquaculture. Ithaca Publishing Company, USA.

Tokunaga, K., Tamaru, C., Ako, H., Leung, P., 2015. Economics of small-scale commercial aquaponics in hawai'i. Journal of the World Aquaculture Society 46 (1), 20–32. Available at https://doi.org/10.1111/jwas.12173.

Tyson, R.V., Treadwel, D.D., Simonne, E.H., 2011. Opportunities and challenges to sustainability in aquaponic systems. Horttechnology 21 (1), 6–13. Available at https://doi.org/10.21273/horttech.21.1.6.

Vassiliou, V., Charalambides, M., Menicou, M., Chartosia, N., Tzen, E., Evagelos, B., et al., 2015. Aquaculture feed management system powered by renewable energy sources: Investment justification. Aquaculture Economics and Management 19 (4), 423–443. Available at https://doi.org/10.1080/13657305.2015.1082115.

Wei, Y., Li, W., An, D., Li, D., Jiao, Y., Wei, Q., 2019. Equipment and intelligent control system in aquaponics: a review. IEEE Access 7, 169306–169326. Available at https://doi.org/10.1109/ACCESS.2019.2953491.

Wirza, R., Nazir, S., 2021. Urban aquaponics farming and cities—a systematic literature review. Reviews on Environmental Health 36 (1), 47–61. Available at https://doi.org/10.1515/reveh-2020-0064.

Yanes, A.R., Martinez, P., Ahmad, R., 2020. Towards automated aquaponics: A review on monitoring, IoT, and smart systems. Journal of Cleaner Production 263. Available at https://doi.org/10.1016/j.jclepro.2020.121571.

Zou, Y., Hu, Z., Zhang, J., Xie, H., Liang, S., Wang, J., et al., 2016. Attempts to improve nitrogen utilization efficiency of aquaponics through nitrifies addition and filler gradation. Environmental Science and Pollution Research 23 (7), 6671–6679. Available at https://doi.org/10.1007/s11356-015-5898-0.

CHAPTER

Case studies of aquaponics systems worldwide

11

Daoliang Li[1,2], Meng Cui[3], Haojie Chang[3] and Yaru Chen[3]

[1]*National Innovation Center for Digital Fishery, China Agricultural University, Beijing, P.R. China,*
[2]*College of Information and Electrical Engineering, China Agricultural University, Beijing, P.R. China,*
[3]*Centre for Vision Speech and Signal Processing, University of Surrey, Guildford, United Kingdom*

11.1 America continents

Globally, aquaponics technology, which combines aquaculture and hydroponics, is rapidly advancing as a sustainable agricultural method. Both North America and South America have made significant strides in this field. In North America, especially in the United States, early experiments and trials conducted by universities and research institutions have promoted the adoption and spread of aquaponics technology. In South America, countries such as Brazil, Argentina, and Chile have actively introduced and adapted aquaponics technology in recent years, developing systems tailored to their local climates and economic conditions.

11.1.1 Integrated Aqua-Vegeculture System

The Integrated Aqua-Vegeculture System (IAVS) developed by North Carolina State University is a milestone in aquaponics technology (McMurtry et al., 1997). Designed by researchers at the university, this system aims to achieve efficient water use and sustainable production by integrating aquaculture and plant cultivation. The IAVS system's main features include efficient water resource utilization, sustainable production methods, and significant economic benefits.

11.1.1.1 Efficient water resource utilization

The core design of the IAVS system is to purify water through plant growth, thereby seamlessly integrating fish farming and plant cultivation. By recycling water resources, the system significantly improves water use efficiency. Specifically, waste produced in fish farming is absorbed by plants, and the purified water is recirculated back to the fish tanks, creating a closed-loop system. This approach not only reduces water waste but also greatly decreases reliance on external water sources.

Smart Aquaponic Farming. DOI: https://doi.org/10.1016/B978-0-443-28945-3.00002-5

11.1.1.2 Sustainable production methods

The IAVS system uses natural biological filtration processes to reduce the need for chemical agents and minimize wastewater discharge, thus lowering environmental impact. By relying on the absorption capacity of plant roots to convert nutrients in the water into essential growth materials, the system avoids the use of chemical fertilizers, achieving sustainable agricultural production. In addition, this natural filtration mechanism effectively purifies the water within the system, providing a healthy environment for fish growth.

11.1.1.3 Significant economic benefits

The IAVS system is capable of producing both fish and vegetables simultaneously, offering high economic returns. This system provides farmers with a diversified source of income, giving them a competitive edge in the market. Through aquaponics technology, farmers can produce high-quality aquatic products and fresh organic vegetables, meeting the market's demand for healthy food. Furthermore, the efficiency and sustainability of the IAVS system make it a highly attractive investment in agricultural production, drawing significant interest from investors and agricultural enterprises.

11.1.1.4 Conclusion

The IAVS system at North Carolina State University stands out for its unique advantages in design, technological innovation, and practical application. By ensuring efficient water resource use, sustainable production methods, and significant economic benefits, the IAVS system lays a solid foundation for the development of modern aquaponics technology. It not only offers an efficient agricultural production method but also promotes the adoption and application of aquaponics technology globally. In the future, the IAVS system is expected to be implemented in more countries and regions, contributing to global food security and environmental protection efforts.

11.1.2 University of the Virgin Islands system

The UVI (University of the Virgin Islands) system is another aquaponics innovation with a profound impact on modern aquaponics development. Known for its efficient design and outstanding production capacity, the UVI system (Rakocy et al., 2006) optimizes water circulation and nutrient supply to achieve synchronous growth of fish and plants. The innovative design and successful application of the UVI system have made it a model for aquaponics worldwide.

11.1.2.1 Optimized water circulation system

The UVI system employs an efficient water circulation system with multistage filtration and aeration equipment to ensure optimal water quality. This design not only accelerates fish growth but also promotes healthy plant growth. Specifically, the water circulation process in the UVI system involves water from the fish tanks passing through mechanical and biological filters before further purification by plant roots, finally returning to the fish tanks. This multilayer filtration mechanism ensures water purity, providing an ideal growth environment for both fish and plants.

11.1.2.2 ***Modular design***

The UVI system is highly flexible and scalable. Farm operators can adjust the system's scale and configuration according to their needs to maximize production efficiency. The modular design of the UVI system allows it to adapt to different production conditions and market demands. Whether for small-scale family farms or large-scale commercial agricultural production, the system can be optimized by adjusting the number and configuration of modules to achieve the best production results. This flexibility has led to the widespread adoption of the UVI system globally.

11.1.2.3 ***Broad application prospects***

The success of the UVI system extends beyond academic research to commercial production. Many countries and regions have adopted the UVI system's design to establish their aquaponics farms. The efficiency and reliability of the UVI system make it a benchmark in aquaponics technology, attracting significant investment and technical collaboration. For example, in the Middle East, where water resources are scarce, the UVI system is widely used in desert agriculture, providing a sustainable agricultural production model. In Asia, particularly China and India, the UVI system has been introduced to enhance agricultural productivity and food safety.

11.1.2.4 ***Conclusion***

The UVI system demonstrates the adaptability and efficiency of aquaponics technology in various environments and conditions through its optimized water circulation system, modular design, and broad application prospects. The UVI system has achieved remarkable results in academic research and shown significant potential in practical applications. By learning from the successful experiences of the UVI system, other countries and regions can adopt its design principles and technological innovations to advance local aquaponics technology, achieve sustainable agricultural development, and environmental protection goals. As technology advances and the global demand for sustainable agriculture increases, the UVI system will continue to play a crucial role worldwide, contributing more to solving food security and environmental protection issues.

11.1.3 Aquaponics farms in Wisconsin

In addition to the above two typical cases, aquaponics farms in Wisconsin, United States, represent another important example of the practical application of aquaponics technology (Goddek et al., 2019). These farms showcase the wide application and development potential of aquaponics in modern agriculture through practical operations and continuous innovation.

11.1.3.1 Background and development

Wisconsin's aquaponics farms are often operated by family farmers and small agricultural businesses. These farms are typically located in urban or suburban areas, producing fresh organic vegetables and high-quality aquatic products through aquaponics systems. This production model not only meets the local residents' demand for healthy food but also provides farmers with a sustainable source of income.

11.1.3.2 System design and technical features

Wisconsin's aquaponics farms employ various system designs and technologies, adapting to local climate conditions and market demands to develop efficient, low-cost production models. The following are some key system design and technical features:

1. Simplified water circulation system: Compared to large commercial systems, Wisconsin farms usually adopt simplified water circulation systems. These systems use basic mechanical and biological filtration techniques to maintain water quality balance through nutrient absorption and purification by plant roots.
2. Local adaptability: Farmers select vegetable varieties and fish species suitable for local climate conditions and market demands. For example, in areas with long winters, they might choose cold-tolerant vegetable varieties and fish species, while in regions with hot summers, they might implement shading and cooling measures to ensure the system's stable operation.
3. Energy and environmental measures: Many Wisconsin aquaponics farms implement energy-saving and environmental measures, such as using solar power, rainwater collection, and recycling. These measures reduce production costs and environmental impact, enhancing the farm's sustainability.

11.1.3.3 Economic benefits and social impact

Wisconsin's aquaponics farms have gained local consumer favor by providing fresh, organic farm products. Through direct sales, farmers' markets, and Community Supported Agriculture programs, these farms not only secure stable income but also strengthen their connection and interaction with the community. In addition, aquaponics farms create local job opportunities and promote rural economic development (Love et al., 2015).

11.1.3.4 Continuous innovation and future development

Wisconsin's aquaponics farms continually innovate and optimize systems, exploring more efficient and sustainable production models. In the future, these farms may further advance aquaponics technology, contributing to local and global sustainable agriculture and environmental protection goals. Through their practical experience and successful application, Wisconsin's aquaponics farms offer valuable insights and models for aquaponics development worldwide.

11.2 European continents

In recent years, the aquaponics industry in Europe has developed rapidly, emerging as a significant innovative force in agriculture and aquaculture. Aquaponics systems combine aquaculture with hydroponics, forming an efficient closed-loop ecosystem. Fish waste is broken down by microorganisms into nutrients that plants can absorb, and the plants purify the water through their root systems, recycling it back to the fish tanks. This model not only enhances resource utilization efficiency but also reduces environmental pollution, making it an effective approach to achieving sustainable agriculture.

11.2.1 Technical features

The technical features of European aquaponics systems have distinct regional characteristics, mainly including the following aspects:

1. **Advanced recirculating water treatment technology:** Europe has significant advantages in water treatment technology. Many aquaponics systems use advanced physical, chemical, and biological filtration techniques to ensure efficient and stable water quality purification.
2. **Energy optimization and renewable energy integration:** Many European aquaponics systems incorporate renewable energy sources such as solar and wind power, reducing the system's energy consumption. For example, some projects in Denmark use geothermal and solar energy to power the systems, achieving low-carbon and environmentally friendly goals.
3. **Precision control and intelligent management:** European aquaponics systems widely use sensor networks and Internet of Things (IoT) technology. Real-time data monitoring and big data analysis are used to precisely control water quality, temperature, light, and nutrients, achieving intelligent management and optimization of the systems.
4. **Multilayer vertical farming systems:** To enhance space utilization, some urban agriculture projects in Europe have developed multilayer vertical farming systems. This design not only saves space but also increases the output per unit area, suitable for the development of aquaponics in urban environments.

5. **Comprehensive waste management:** Many European aquaponics systems emphasize comprehensive waste management and recycling, including using fish waste for composting, which is then used for plant cultivation, enhancing overall resource utilization efficiency (Palm et al., 2018).

11.2.2 Industry characteristics

The characteristics of the European aquaponics industry lie in its significant environmental and economic benefits:

1. **Environmentally friendly:** Aquaponics systems significantly reduce the use of fertilizers and pesticides, lowering soil and water pollution, thus contributing to environmental protection.
2. **Policy support:** European governments highly value aquaponics systems, promoting the technology through policy support and funding. For example, EU research projects and funding support provide important guarantees for technological innovation and industry standardization.
3. **Technological innovation:** Europe has significant investments in innovation in aquaponics technology (Körner et al., 2021). Countries such as the Netherlands and Germany have made notable progress in developing new recirculating systems, optimizing nutrient ratios, and efficient water treatment technologies.
4. **Education and training:** European universities and research institutions conduct extensive education and training programs in aquaponics (Karimanzira and Jacobi, 2022), from basic research to commercial applications, promoting technology dissemination and talent cultivation.
5. **Urban agriculture:** Many European cities, such as London, Berlin, and Copenhagen, actively promote urban aquaponics projects. These projects not only enhance the self-sufficiency of urban food supplies but also increase public awareness and participation in sustainable agriculture.
6. **Market acceptance:** European consumers generally have a high recognition of organic and environmentally friendly foods, which provides favorable conditions for the market promotion of aquaponics products.

11.2.3 Regional differences

There are certain differences in the maturity and application of aquaponics technology across European countries:

1. **Technological leadership:** Western European countries such as the Netherlands and Germany (Körner et al., 2021) lead in technology development and commercial application, with mature commercial aquaponics farms and comprehensive technical support systems.

2. **Development potential:** Eastern European countries are in the early stages of developing the aquaponics industry but show great potential for growth with the spread of technology and policy support.
3. **Market acceptance:** Western European consumers generally have a high acceptance of environmentally friendly and organic foods, making it easier to promote aquaponics products in the market.

11.2.4 Market prospects

As environmental awareness and healthy eating concepts spread, the acceptance of aquaponics systems in the European market is gradually increasing. Governments and research institutions in various countries are actively promoting the technology through policy support and funding. In the future, with continuous technological advancements and market maturity, aquaponics systems are expected to occupy a significant position in European agriculture, becoming a key force in promoting sustainable agricultural development.

11.2.5 Conclusion

In summary, the aquaponics industry in Europe demonstrates strong advantages in technological innovation, policy support, market acceptance, and education and training. In the following chapters, we will explore specific case studies to delve into the practical applications and future prospects of this industry, providing readers with a comprehensive and in-depth understanding.

11.2.5.1 Les Nouvelles Fermes

In Bordeaux, France, an innovative agricultural company is transforming the future of urban farming. Les Nouvelles Fermes, with its advanced aquaponics systems and information technology, has become a model of modern sustainable agriculture. Since its establishment, Les Nouvelles Fermes has rapidly expanded from its initial experimental farm "Pauline" to the soon-to-launch large-scale urban farm "Odette," showcasing its ambitious vision for urban agriculture.

11.2.5.1.1 Experimental farm "Pauline"

The "Pauline" experimental farm, located in the Lormont area of Bordeaux, covers 1000 square meters and produces 20 tons of fresh vegetables and 2 tons of rainbow trout annually, validating the feasibility of the aquaponics model. The success of this farm provided Les Nouvelles Fermes with valuable experience and data, helping them continuously optimize their technology and prepare for larger-scale projects.

11.2.5.1.2 Large-scale urban farm "Odette"

To achieve greater production scale and social impact, Les Nouvelles Fermes plans to launch the "Odette" urban farm by the end of 2021. This new farm, located in Mérignac, Bordeaux, covers 5000 square meters, will create 17 new jobs, and significantly increase production capacity. The establishment of the "Odette" farm represents a critical step toward large-scale production for Les Nouvelles Fermes, aiming to provide fresh food to more urban residents.

The "Odette" urban farm has clear production goals: to produce 60 tons of fresh vegetables and 12 tons of rainbow trout annually. By employing efficient aquaponics systems and information management technologies, the farm aims to significantly improve production efficiency and environmental benefits. Utilizing advanced water treatment and resource recycling technologies, the "Odette" farm is expected to reduce water consumption by 90%–95% and energy consumption by 4–5 times compared to traditional agriculture. The system's production efficiency is also greatly enhanced, with certain crops like Swiss chard achieving yields of up to 24 kg/m²/year, compared to 2–3 kg/m²/year in traditional open-field cultivation.

Les Nouvelles Fermes employs various advanced information technologies in its aquaponics systems to ensure efficient and sustainable production. The company integrates highly automated control and monitoring systems, utilizing sensor networks and IoT technology to monitor and adjust key parameters in real time (Fig. 11.1). Sensors continuously monitor water pH, dissolved oxygen, ammonia, and nitrate levels to maintain water quality suitable for both fish and plant growth. Temperature sensors monitor ambient and water temperatures, automatically adjusting heating or cooling equipment to maintain optimal growth conditions.

FIGURE 11.1

Indoor breeding greenhouse.

Light sensors monitor the intensity of light in the greenhouse, supplemented by LED lighting systems to provide the necessary light for plant growth. These measures ensure a stable environment, promoting the healthy growth of plants and fish.

The company leverages big data analysis to optimize production processes and management decisions. Growth data is recorded to build a detailed database, providing scientific support for optimizing breeding and planting plans. Through data analysis and machine learning, the system can predict potential issues and take preventive measures in advance, enhancing system stability and production efficiency. In addition, Les Nouvelles Fermes employs efficient water treatment technologies to ensure the recycling of water resources and the sustainability of the system. Biological filtration utilizes microorganisms to convert harmful substances in fish waste into nutrients that plants can absorb. Mechanical filtration removes suspended particles and solid waste from the water, and UV disinfection eliminates pathogens and algae, ensuring water safety.

To maximize space utilization, Les Nouvelles Fermes employs multilayer vertical farming systems. In limited urban spaces, multilayer planting beds are established to maximize vertical space use and increase yield per unit area. The smart system manages water flow, nutrients, and light for each layer, ensuring optimal growth conditions for all plants. In addition, Les Nouvelles Fermes is committed to reducing energy consumption and actively incorporates renewable energy. Solar panels are installed on the roofs of greenhouses and aquaculture facilities to provide clean energy for the system, and the energy management system optimizes energy use across devices to reduce waste and improve efficiency. The company also emphasizes comprehensive waste management and recycling, composting organic waste from fish and plant production to create fertilizer for further plant cultivation.

Les Nouvelles Fermes exemplifies the potential of future urban agriculture through its advanced technology and sustainable models. The extensive application of information technology not only improves production efficiency but also significantly reduces environmental impact, making it a successful model of modern urban agriculture. As Les Nouvelles Fermes scales up its operations, it aims to promote its innovative model globally, contributing to sustainable agricultural development.

11.2.5.2 Building integrated greenhouses

The skyline of Brussels bears witness to the rise of an innovative agricultural company: BIGH (Building Integrated Greenhouses). BIGH is dedicated to integrating greenhouses with urban buildings and promoting sustainable urban agriculture through unique aquaponics systems. Their flagship project is located on the roof of the Anderlecht slaughterhouse in Brussels, making it one of the largest urban aquaponics farms in Europe.

FIGURE 11.2

The Anderlecht slaughterhouse rooftop project.

BIGH's Anderlecht slaughterhouse rooftop project (Fig. 11.2) covers 4000 square meters, including 2000 square meters of high-tech greenhouses and 2000 square meters of outdoor planting areas. Combining advanced greenhouse technology with circular economy principles, BIGH produces several tons of organic vegetables and fish annually. These products are directly supplied to restaurants and retailers within Brussels, significantly reducing transportation distances and associated carbon emissions. BIGH employs highly automated and information-driven technologies to ensure efficient and sustainable production. The company integrates automated control and monitoring systems, using sensor networks and IoT technology to monitor and adjust key parameters such as water quality, temperature, and light intensity in real-time. These technologies ensure optimal growth conditions, promoting the healthy growth of both fish and plants.

BIGH uses solar panels to power the system and a rainwater collection system to mix rainwater with fishpond water for plant irrigation. The company also employs heat pump technology to recover heat from buildings to manage the temperature in greenhouses and fish ponds, minimizing energy consumption. In addition, BIGH focuses on using renewable resources and recyclable materials. Building materials are made as much as possible from recyclable and renewable resources, such as plant fiber walls and recycled panels. They minimize excess packaging by prioritizing recycled materials and cardboard, reducing environmental impact.

Since its establishment, BIGH has expanded its innovative model and technologies across multiple projects in Brussels and other European cities. The Anderlecht slaughterhouse project alone produces 35 tons of fish, 15 tons of tomatoes, and 170,000 pots of kitchen herbs annually (Fig. 11.3), demonstrating

FIGURE 11.3

The fish pond at the slaughterhouse.

their strong production capacity and market demand. Through the sale of these high-quality products, the company has achieved significant economic benefits. BIGH's projects not only provide fresh, healthy food to urban residents but also raise public awareness of sustainable agriculture through education and community activities. They collaborate with local communities and schools to conduct agricultural education programs, aiming to promote sustainable agricultural practices and foster urban agriculture development.

BIGH collaborates with schools and community organizations to offer educational projects and workshops, spreading knowledge about sustainable agriculture and inspiring public interest and participation in urban farming. By reducing transportation distances and carbon emissions, BIGH significantly lowers its environmental footprint. Their efficient resource management system reduces water and energy consumption, providing a sustainable food production model for cities. BIGH plans to expand its model to more cities in the coming years, continuously innovating and optimizing technologies to further enhance the sustainability and efficiency of urban agriculture. The company aims to use its advanced technology and model to drive global urban agriculture development, contributing to sustainable urban living.

BIGH is exploring opportunities to extend its successful model to other international cities, building more urban greenhouse projects through partnerships and investments, thus expanding its influence. The company will continue to invest in the research and application of new technologies, constantly improving automation and resource management to ensure the efficient operation and environmental sustainability of its urban agriculture projects. Through its innovative urban agriculture model and advanced technology, BIGH showcases the enormous potential of future urban farming, becoming a successful example of modern urban agriculture. As BIGH scales up its operations, it is poised to promote its innovative model globally, contributing to the vision of sustainable agricultural development.

11.3 Asia continents

Aquaponics, the combination of aquaculture and hydroponics, is transforming sustainable farming across Asia. This integrated system maximizes resource efficiency and supports environmentally friendly agricultural practices, offering solutions to food security, resource conservation, and economic development. In the following sections, we will explore several cases of aquaponics in Asia, highlighting their approaches and illustrating benefits they bring to the region.

11.3.1 Xinjin aquaponics farm: large scale aquaponics in Si Chuan

Located in Xinjin County, Chengdu, YoGoTime Farm (Tianfu Agricultural Expo Park, 2025) stands out as a leading example of aquaponics in China, where technology meets traditional farming. The farm spans 30 acres and includes 28 greenhouses, producing about 50 tons of fish and vegetables each year, primarily for tourists (Fig. 11.4).

YoGoTime Farm utilizes an innovative cycle called "fish-manure water to clean-vegetable water to aquaculture." In this system, waste from over eight types of freshwater fish feeds more than 50 varieties of vegetables and fruits. The plants

FIGURE 11.4

An large aquaponics farm in Xinjin.

FIGURE 11.5

Crops in the farm.

clean the water, which is then reused in the fish tanks. Special bacteria help break down the fish waste, maintaining the system's balance and sustainability. The farm's approach brings numerous ecological benefits, such as water conservation and the elimination of chemical fertilizers, which in turn reduces environmental pollution. Economically, the farm thrives by selling products directly to consumers online, which increases profits and showcases the farm as a model of modern agricultural practices (Fig. 11.5).

Socially, the farm is a significant local employer and promotes aquaponics through training programs. It engages the community with agricultural tourism and educational tours, enhancing awareness and support for sustainable agriculture.

YoGoTime Farm also contributes to food safety by providing clean, residue-free produce. The founder, Chen Bin, shifted from software development to agriculture, committed to sustainable farming. He translated extensive aquaponics research to adapt the technology to local conditions. Despite challenges like fish diseases and equipment failures, his dedication has made the farm resilient and innovative. Looking ahead, Chen aims to turn the farm into a creative agricultural theme park to educate young people about sustainable farming. The farm is not just a successful aquaponics operation in China. It is also a symbol of how modern agriculture can drive rural revitalization and sustainable development.

11.3.2 Full nature farms: revolutionizing aquaponics in Hong Kong

Located in the industrial districts of Tsuen Wan and Kwai Chung, Full Nature HK Farms (Full Nature Farms, 2025) is at the forefront of integrating aquaponics with cutting-edge agricultural technologies in Hong Kong. The farm, initially established in a 25,000 square foot greenhouse, has adeptly grown over 500 types of crops through aquaponics—a method that merges fish farming with hydroponics. In this system, fish waste provides nutrients for plants, which in turn purify the water for the fish, crafting a sustainable ecosystem.

FIGURE 11.6

Vertical farm techniques for aquaponics.

Given the premium on space in Hong Kong, Full Nature Farms employs vertical farming techniques to enhance space efficiency and crop yield per square foot. This approach not only leverages vertical space but also significantly boosts production (Fig. 11.6).

The farm incorporates various advanced technologies to maintain optimal conditions for both fish and plants. It boasts a sophisticated monitoring system that controls temperature, humidity, and pH levels, essential for the health of the ecosystem. Leveraging artificial intelligence and the IoT, the farm simulates soil environments and decomposes organic materials to enhance the nutrient content and flavor of the vegetables (Fig. 11.7).

A proprietary AI-driven control system stands as the highlight of the farm's technological advancements, allowing for precise environmental adjustments and real-time monitoring. This innovation not only shortens the vegetable growth cycle but also elevates their quality.

Full Nature Farms' aquaponics system greatly minimizes the need for water and chemical fertilizers, a crucial advantage in urban areas focused on sustainability. The farm doubles as an educational hub, promoting "Farm to Table" experiences to boost public engagement and awareness of sustainable practices. The farm has attracted attention from local five-star hotels and high-end restaurants, ensuring a

FIGURE 11.7

AI-driven control system.

consistent demand for its produce. It is also venturing into retail with boxed salads and plans to export its indoor farming technology across Hong Kong.

Full Nature Farms positions itself as a high-end brand in the market, promoting its commitment to organic, pesticide-free farming, zero emissions, and no external pollution. This approach perfectly meets the demands of the premium market, supplying fish and vegetables to five-star hotels and upscale restaurants.

11.3.3 Aquaponics Hub Asia

In Southeast Asia, innovative aquaponics systems are reshaping sustainable agriculture. Based in Khao Mai Kaeo, in Thailand's Bang Lamung District, Aquaponics Hub Asia (AHA, 2025) exemplifies how technology can enhance traditional farming practices in Thailand and Malaysia. AHA is at the forefront of implementing both domestic and large-scale commercial aquaponics systems in the region. Founded to transform sustainable farming, AHA has become a leading Fig. 11.8 in aquaponics, enhancing food security and promoting environmental stewardship.

The heart of AHA's operations lies in the symbiotic relationship between fish and plants, fundamental to aquaponics. This system merges aquaculture (raising fish) with

FIGURE 11.8

Commercial units and domestic units.

hydroponics (growing plants soil-free), where fish waste provides nutrients for plants. The plants, in turn, clean the water that cycles back to the fish tanks. This efficient loop curtails the need for water and chemical fertilizers, offering a perfect solution for areas experiencing water scarcity and land degradation. AHA's facilities in Thailand and Malaysia incorporate cutting-edge technologies to enhance growth conditions for aquatic and plant life. Through vertical farming and LED lighting, AHA boosts space utilization and crop yield—critical in densely populated regions. The farms produce diverse crops and fish species, meeting local preferences and market demands, ensuring the production systems' relevance and sustainability.

AHA significantly bolsters local economies by reducing import reliance and bolstering self-sufficiency. The operations also create jobs and offer training for local communities, boosting livelihoods and fostering inclusive growth. AHA is poised to extend its influence throughout Southeast Asia and beyond, seeking partnerships to transport its technology and expertise to new markets. This includes potential urban projects where space is limited, and sustainability is paramount.

Aquaponics Hub Asia is more than an agricultural business—it is a beacon of how innovation aligned with sustainability can create substantial benefits. Its advanced systems epitomize agricultural innovation in Asia, blending economic viability with ecological sensitivity to pave the way toward a sustainable, food-secure future. By leveraging its strategic position in Thailand and extensive operations in Malaysia, AHA is poised to continue leading the aquaponics sector, driving a green revolution in Asian agriculture.

11.3.4 Japan—Akponi Co

In Aisai City, Aichi Prefecture, Japan, the Tsunagu Farm (Horti Daily, 2023) stands as a beacon of innovation in aquaponics, merging advanced hydroponic techniques with aquaculture to create a sustainable model of urban farming. This farm, one of the largest of its kind in Japan, spans 900 square meters and has the capacity to produce approximately 20,000 plants.

Tsunagu Farm employs a vertical hydroponic system that significantly enhances planting density and spatial efficiency. This system supports the cultivation of leafy vegetables like lettuce and Italian parsley, alongside aquaculture of tilapia, a fish widely used in local cuisine. The integration of these species exemplifies a balanced ecosystem where plant and fish life support one another, with the plants benefiting from the nutrients provided by fish waste, and the fish in turn thriving in the clean water filtered by the plants (Fig. 11.9).

The farm is equipped with patented technology that allows for independent operation of the aquaculture and hydroponic components in the event of issues, ensuring continuous production and supply. This level of technological advancement is pivotal in maintaining the delicate balance of the aquaponics system and in optimizing both growth and yield. Resource efficiency is a hallmark of Tsunagu Farm's operation. The system conserves up to 80% more water than traditional farming methods, while also eschewing the use of pesticides and chemical fertilizers. These practices not only sustain environmental health but also produce clean, organic food products that are increasingly demanded by health-conscious consumers.

Economically, the farm focuses on local consumption, which reduces transportation costs and carbon footprints, and uses its innovative vertical farming technology to maximize yield per unit area. This approach not only supports the local economy but also sets a benchmark for profitable urban farming. Social impact is deeply integrated into Tsunagu Farm's operations. In collaboration with local organizations, the farm provides employment opportunities for disabled individuals, integrating

FIGURE 11.9

Vertical hydroponic system.

FIGURE 11.10

Fish units.

social welfare into its business model. This inclusivity enhances the community's economic diversity and promotes social cohesion. Tsunagu Farm also embraces sustainability through innovative practices. The use of refurbished greenhouses minimizes initial costs and promotes resource reuse. This strategy serves as a model for revitalizing abandoned farmlands and demonstrates how sustainable practices can lead to economic and ecological benefits (Fig. 11.10).

As one of the foremost aquaponics farms in Japan, Tsunagu Farm represents a successful amalgamation of advanced hydroponic technology with social welfare programs, creating a sustainable and scalable model of agriculture that supports both community development and environmental stewardship. Through its operations, Tsunagu Farm illustrates the profound potential of integrated farming systems in urban settings, providing a blueprint for future agricultural innovations.

11.4 Australian continents

Australia has been at the forefront of adopting aquaponics, a sustainable food production system that combines aquaculture (raising fish) with hydroponics (growing plants without soil). This innovative approach has gained traction across the country, with numerous projects and businesses showcasing its potential, particularly in urban settings like backyards and courtyards. One notable example is Biofilta (2025), a Melbourne-based urban farming company that has revolutionized the way we think about growing food in cities.

Biofilta has developed a modular, scalable, and water-efficient aquaponics system called the "Food Cube." These compact units can be easily installed on rooftops, in backyards, or even inside buildings, making them perfect for urban environments where space is limited. The Food Cube operates using a flood and drain system, where nutrient-rich water from the fish tank is pumped into the growing beds, nourishing the plants before draining back into the tank. This closed-loop system reduces water consumption by up to 90% compared to traditional agriculture, while also eliminating the need for chemical fertilizers and pesticides (Fig. 11.11).

The environmental benefits of aquaponics are significant, especially in a country like Australia where water scarcity is a major concern. By recirculating water from the fish tank, aquaponics minimizes water loss through evaporation, percolation, or runoff, unlike conventional farming methods. In addition, the use of chemicals is dramatically reduced, as fish waste provides the necessary nutrients for plant growth. This not only benefits the environment but also ensures a healthier and more sustainable food production system.

Aquaponics also helps mitigate the problem of soil erosion, which is a significant issue in Australia, causing the loss of precious topsoil and silt build-up in waterways. As aquaponics does not require ploughing or tilling of the soil, it effectively reduces erosion. Furthermore, when set up in a controlled environment like a greenhouse, aquaponics reduces the need for pesticides, as the likelihood of pest infiltration is minimized.

From a commercial perspective, aquaponics offers several advantages for producers. In Australia, high-quality agricultural land with fertile soil and abundant water access is scarce and expensive. Aquaponics provides an alternative, allowing food production to take place in various locations, as long as there is a water supply,

FIGURE 11.11

John Bowskill's greenhouse in Witchcliffe is among the largest aquaponics set-ups in WA.

which can be as simple as rainwater stored in tanks. This opens up opportunities for individuals to grow their own food, regardless of their location or access to prime agricultural land.

Moreover, aquaponics can help reduce the running costs associated with conventional horticulture. Traditional farming often requires expensive equipment like tractors and implements, as well as significant expenditure on synthetic chemicals and fuel. In contrast, aquaponic systems eliminate the need for ploughing, spraying, and heavy machinery, thereby reducing operational costs and fossil fuel consumption.

The success of companies like Biofilta highlights the growing interest in aquaponics as a sustainable solution to the challenges faced by modern agriculture. As more Australians embrace this technology, particularly in urban settings like backyards and courtyards, we can expect to see a greater variety of fresh, locally-grown produce in our cities. This shift toward sustainable, local food production will not only improve food security and community health but also contribute to a more resilient and environmentally friendly food system overall.

11.5 Africa continents

Aquaponics, an innovative food production system that combines fish farming with soilless plant cultivation, is emerging as a promising solution to address Africa's complex food security and environmental challenges. As the continent grapples with rapid population growth, climate change, and resource scarcity, aquaponics offers a sustainable and efficient way to produce nutritious food while conserving water and land resources (Mchunu et al., 2018b).

One of the most significant advantages of aquaponics in the African context is its potential to alleviate poverty and create sustainable livelihoods, particularly in rural and peri-urban areas. By integrating fish farming with vegetable production, smallholder farmers can diversify their income streams, reduce their reliance on rain-fed agriculture, and increase their resilience to economic and environmental shocks. This is particularly important in regions like sub-Saharan Africa, where the majority of the population depends on agriculture for their livelihoods, yet face significant challenges such as soil degradation, water scarcity, and limited access to markets.

In countries like Kenya, Nigeria, and South Africa (Mchunu et al., 2018a), a growing number of aquaponics projects are being developed to empower smallholder farmers and strengthen local food systems. For example, in Kenya's Kibera slum, the largest urban slum in Africa, a community-based organization called Kibera Fresh has established an aquaponics farm that provides fresh vegetables and fish to local residents while also creating employment opportunities for youth and women. Using locally available materials and adapting the technology to the local context, Kibera Fresh demonstrates the potential of aquaponics to transform urban food systems and improve the livelihoods of marginalized communities.

FIGURE 11.12

At Stellenbosch University's Welgevallen aquaponics farm, herbs like basil are produced for local restaurants.

Another key aspect of aquaponics development in Africa is the focus on capacity building and knowledge transfer. Many projects prioritize training and education to equip farmers, students, and entrepreneurs with the skills and knowledge needed to design, operate, and scale aquaponics systems. For instance, the West African Aquaponics Research and Education Centre, based in Ghana, offers practical training courses and workshops to promote aquaponics as a viable agribusiness model and foster innovation in the sector.

Furthermore, aquaponics is also being explored as a means to enhance nutrition and food diversity in Africa, particularly in regions where access to animal protein and fresh produce is limited (FAO, 2014). By providing a locally sourced, year-round supply of fish and vegetables, aquaponics can help combat malnutrition and improve the health outcomes of vulnerable populations, such as children and pregnant women.

Despite the growing interest in aquaponics across Africa, the sector still faces significant challenges, including limited access to finance, infrastructure, and policy support. To unlock the full potential of aquaponics for sustainable food production and rural development, there is a need for greater investment in research and development, market linkages, and enabling policies that recognize the unique role of aquaponics in Africa's food systems.

As African countries seek to build more resilient, equitable, and sustainable food systems, aquaponics offers a promising pathway to address the continent's pressing food security and environmental challenges. By harnessing the power of innovation, collaboration, and local knowledge, Africa can position itself at the forefront of the global aquaponics movement and pioneer new models for sustainable food production that benefit both people and the planet (Fig. 11.12).

References

AHA. (2025). AHAponics process. Available at https://aha-services.com/aquaponics.

Biofilta, (2025). What is the urban agriculture forum? Available at https://uaf.org.au/about/partners/biofilta/.

FAO. (2014). *Small-scale aquaponic food production: Integrated fish and plant farming*. FAO Fisheries and Aquaculture Technical Paper No. 589. Food and Agriculture Organization of the United Nations.

Full nature farms, (2025). Transforming local agriculture through sustainable agritech innovation. Available at https://fullnature.com/urban-oasis-lab.

Goddek, S., Joyce, A., Kotzen, B., Burnell, G.M. (Eds.), 2019. Aquaponics Food Production Systems: Combined Aquaculture and Plant Agriculture Production Strategies for the Future. Springer Nature.

Horti daily, (2023, March 30). Japan: new hybrid aquaponic farm 'to be largest yet'. Available at https://www.hortidaily.com/article/9515428/japan-new-hybrid-aquaponic-farm-to-be-largest-yet.

Karimanzira, D., Jacobi, M., 2022. Digital twin-based control and monitoring of an aquaponics system. Computers and Electronics in Agriculture 197, 106967.

Körner, O., Holst, N., Zwart, H.F.D., 2021. Energy saving potential of climate control in greenhouses. Agricultural Systems 190, 103061.
Love, D.C., Fry, J.P., Genello, L., Hill, K.E., Frederick, J.A., Li, X., et al., 2015. An international survey of aquaponics practitioners. PLoS One 10 (7), e0133647.
Mchunu, N., Lagerwall, G., Senzanje, A., 2018a. Aquaponics in South Africa: results of a national survey. Aquaculture Reports Volume 12, 12–19.
Mchunu, N., Lagerwall, G., Senzanje, A., 2018b. Aquaponics in South Africa: status, challenges and opportunities. Water SA 44 (3), 353–363.
McMurtry, M.R., Sanders, D.C., Patterson, R.P., Nash, A., 1997. Yield of tomato irrigated with recirculatory aquaculture water. Applied Engineering in Agriculture 13 (1), 121–126.
Palm, H.W., Knaus, U., Appelbaum, S., Goddek, S., Strauch, S.M., Vermeulen, T., et al., 2018. Towards commercial aquaponics: a review of systems, designs, scales and components. Fisheries and Aquatic Sciences 21 (1), 1–16.
Rakocy, J.E., Masser, M.P., Losordo, T.M., 2006. Recirculating aquaculture tank production systems: aquaponics—integrating fish and plant culture. SRAC Publication (No), 454.
Tianfu Agricultural Expo Park. (2025, May 14). YOGO farm: Aquaponics Farm, Available at https://www.xinjin.gov.cn/xjxrmzf/c143166/2025/05/14/content_43b2f3daa3df4bd3be8c6403f410ca18.shtml.

Index

Note: Page numbers followed by "*f*" indicate figures and "*t*" indicate tables.

A

T

U

V

W

X

Y

Z

9780443289453